Maturity Model for Effective Management of Construction Teams

TAWFIG KHIDIR IBNOUF

BSc Civil Engineering

MSc, Ph.D. Project Management

Table Of Contents

Chapter 1: Examining The Effective
Management Of Construction Teams 05

Chapter 2: My Research Approach
And Participants 19

Chapter 3: Evaluating My Data And Model 29

Chapter 4: Organizational Culture 47

Chapter 5: Team Composition 59

Chapter 6: Effective Team Building Overview 71

Chapter 7: Managing The Team Life Cycle 81

Chapter 8: Effective Team Management
And Training 91

Chapter 9: Team Leadership 97

Chapter 10: Conclusion 135

References 141

Chapter 1

Examining The Effective Management Of Construction Teams

The construction industry is a dynamic and multifaceted sector, characterized by complex projects that demand effective management of resources, timelines, and teams. Within this context, the management of construction teams emerges as a critical aspect influencing project outcomes. The success of construction projects hinges upon the ability of teams to collaborate efficiently, communicate effectively, and adapt to changing circumstances. However, achieving and maintaining high levels of team performance poses significant challenges for industry professionals.

As a managerial approach, team management focuses on the development and operation of teams. The application of team management principles is instrumental in enhancing employee productivity, promoting self-organization and self-management through collaborative activities, mutual oversight, assistance, and interchangeability. It also involves harnessing individual and group potentials, fostering a shared understanding of values and goals that guide each team member's behavior, and fostering collective responsibility for performance outcomes. Thus, "the success of business organizations and institutions is closely tied to their ability to establish cooperative work environments and effectively manage teamwork" (Zubanov et al., 2017, p. 68). Consequently, organizations have recognized the significance of team management in addressing various challenges that could impede their progress and development.

While the concept of teamwork in the construction industry is not novel, the industry continues to grapple with several team management challenges. Competing relationships and a lack of collaboration are commonplace in the construction sector. Consequently, numerous factors, such as a "lack of trust and ineffective communication pose significant challenges within the construction industry" (Faris et al., 2022). Johnson and Babu (2018) have connected these challenges in construction projects to the need to "constantly deal with uncertainties and temporary teams composed of individuals from diverse linguistic and cultural backgrounds" (p. 514). As a result, these challenges and difficulties have far-reaching consequences on project success, organizational progress, and the construction industry's crucial role in the economy and human development.

The dynamic and complex nature of the construction industry can lead to poor performance and the occurrence of numerous risks that negatively affect project outcomes. Tahmasebinia and Song (2022) identified several factors that affect successful project delivery, including worker conflicts, lack of skilled labor, inexperienced teams, insufficient communication and cooperation among stakeholders, and low labor productivity. Such challenges "affect all parties involved including the client, consultant, and contractors. The impact of how badly it affects the project as a whole depends on the degree of the barriers or challenges to the team characteristic in question" (Dansoh et al., 2017, p. 122). Therefore, the effective management of the construction team becomes a crucial demand for the success of construction projects and the industry as a whole.

Effective teams are promoted by a good organizational culture that includes shared values and common purpose, good teams support, a reliable and motivating rewards and recognition system for proper motivation, good teams development and commitment, a supportive work environment, appropriate selection criteria, ethical work behaviors, and good, well-communicated strategic and operational goals.

The literature has explored project team management from the perspectives of project manager competency, leadership, and effective team characteristics. However, a gap in the literature regards comprehensive methodologies tailored specifically to address the unique challenges faced by construction teams. Research often lacks a holistic perspective, focusing on isolated aspects of team dynamics or management strategies.

The significance here is the potential to introduce novel methods, methodologies, and skills essential for construction team management. Doing so should benefit the construction sector by addressing prevalent challenges such as cost overruns, schedule delays, stakeholder dissatisfaction, and quality issues. It is imperative to recognize that the success or failure of construction projects directly affects both society and local communities. Therefore, successful projects contribute to the well-being of the societies, whereas failures are typically detrimental. Delays in schedule, cost overruns, or stakeholder dissatisfaction can significantly harm these communities. Therefore, the successful management of construction teams becomes paramount for project success, underscoring the core objective of this research.

Countless books, articles, white papers, and seminars have offered a variety of teamwork and team effectiveness models. These models delineate and elucidate the components contributing to successful teamwork. Here, I explore various team effectiveness models to identify the key factors that underpin effective teamwork. This, in turn, will facilitate the development of a robust team management model tailored for construction projects, which is this book's ultimate goal.

Strode et al. (2022) constructed a comprehensive model for enhancing teamwork, drawing upon the "Big Five" model of effective teamwork initially proposed by Salas et al. (2008). The resulting model encompasses shared leadership, team orientation, redundancy, adaptability, and peer feedback as its core components. To facilitate these components, coordinating mechanisms are required, which include shared mental models, effective communication, and mutual trust. Both the developed model and Salas et al.'s model outline the typical behaviors that effective teams exhibit in relation to each component and coordinating mechanism. It is worth noting that while this model was specifically designed for agile teams, the authors acknowledged that its generalizability to all contexts may be limited. However, this limitation is somewhat mitigated by the fact that the model was supported by evidence from two distinct empirical studies that employed different research methodologies and collected data from various countries and continents.

Chidambaram et al. (2021) employed a structured questionnaire to examine project team integration in the construction industry with the aim of formulating a model for promoting effective teamwork within this sector. This model was constructed based on an extensive review of prior literature, with the primary goal of comprehending how team integration impacts project performance and outcomes. The model encompasses several critical dimensions, including team skills, team climate and identity, team structure and situation monitoring, team communication, and team leadership. The authors assert that their study furnishes empirical evidence supporting the notion that the effectiveness of a construction project team significantly influences the project's ultimate outcome. Consequently, they contend that within the construction sector, project managers should exhibit a high degree of sensitivity to the different teams under their supervision or create mechanisms that allow for the expression of concerns, as these actions can positively impact the project's overall results.

Turner et al. (2020) conducted a comprehensive systematic review of various team effectiveness frameworks and models spanning 1998–2018. The authors delineated four fundamental components of team effectiveness (TE), namely: teamwork (TW), task work (TK), outcome or performance (PF), and the value delivered to the customer (CV). This led to the formulation of the following team effectiveness formula: TE = TW + TK + PF + CV.

Furthermore, Turner et al. (2020) incorporated the multidimensional processes inherent in team dynamics as identified by Marks et al. (2001). These multidimensional team processes were categorized into three distinct dimensions: transition phase processes (comprising mission analysis formulation and planning, goal specification, and strategy formation), action phase processes (encompassing progress monitoring toward goals, systems monitoring, team monitoring, backup responses, coordination activities), and interpersonal processes (encompassing conflict management, motivation, confidence building, and effect management). This integration resulted in an expanded formula for team effectiveness, composed of teamwork (TW), the interpersonal phase (IP) processes, task work (TK) as a composite of transition (TP) and action phase (AP) processes, performance (PF) involving quality, quantity, and time-related characteristics, and customer value (CV), which pertains to interactions with external stakeholders and team member experiences. This led to the team effectiveness expanded formula: TE = (TW + IP) + TK (TP + AP) + PF + CV

Turner et al. (2020) argued that

> The transition and action phase processes influence the task work component of team effectiveness, while the interpersonal phase processes impact the teamwork component. Elements employed to assess customer value

> and performance underscore the importance of team member interactions, including team member commitment, satisfaction, and timely task completion. (p. 8)

Latif and Williams (2017) broadened the scope of investigation beyond the conventional examination of attitudes, behaviors, and performance indicators by introducing team formation as a dimension of project team effectiveness. The seven identified dimensions of project team effectiveness encompass team coordination, community social linkage, team performance, knowledge, skills, and attitudes, leadership communication and engagement, decision-making and information sharing, and team formation. Each dimension within this model was characterized by several specific indicators.

The dimension of team coordination assesses behaviors that contribute to team effectiveness, including proactiveness, unity, conflict resolution, information sharing, mutual understanding, responsiveness, learning, and continuous improvement. The factor analysis results revealed that team members' accommodation of each other's needs exhibited the strongest association with the team behaviors factor, followed by the behavior of team members themselves.

The community/social linkage dimension aims to evaluate the effectiveness of the team's connection with the wider society or community. It encompasses integration, contributions to society, conflict resolution, knowledge exchange, respect, network building, addressing issues, and overcoming challenges. The factor analysis results indicated that the item reflecting team members' respect for the community and its culture had the strongest association with the community social linkage factor, followed by the team's establishment of a network within the community. This dimension underscores the significance of considering the team's context in assessing team effectiveness, with the model originating from projects undertaken by nongovernmental organizations with a social focus.

Team performance is another dimension that measures the team's performance through indicators like meeting targets, community satisfaction, responsible resource utilization, turnover during the project, resource loss, work overload, and the impact on society. The factor analysis results showed that the item indicating the satisfaction of the society for which the project was intended exhibited the strongest association with the community social linkage factor, followed by the team's ability to achieve its targets within the specified time frame.

The dimension of knowledge, skills, and attitudes is evaluated using indicators such as skillsets, sincerity, potential, belief in healthy competition, responsibility/accountability, coordination, and mutual respect. The factor analysis results demonstrated that the item indicating mutual respect among team members had the strongest association, followed by team members' belief in fostering healthy competition.

The leader's communication and engagement dimension assesses the effectiveness of the team leader during the project. Effectiveness was measured through indicators such as motivation, continuous follow-up, seeking opinions and facilitating discussions, connecting with the society or community, and the delegation of authority.

The decision-making and information-sharing dimension is intended to measure the effectiveness of decision-making processes and the sharing of information within the team. It is gauged through factors like input, sharing knowledge and experiences, reaching a consensus on issues, sharing ideas openly, and active participation.

Lastly, the team formation dimension's indicators encompass clarity of direction and focus, job understanding, team diversity, and the establishment of realistic goals.

Azmy (2012) developed a model that offers insights into team effectiveness within construction teams, particularly concerning project performance aspects. This model elucidates the variability of team effectiveness factors and equips project team members with a better understanding of the extent to which these factors contribute to project performance. Furthermore, the study produced two distinct definitions of team effectiveness, one from the perspective of construction project team members and another from that of construction project owners. Azmy contended that the team effectiveness factors identified through the research exerted a significant influence on the performance of construction projects, specifically with regard to project change management. These factors encompassed team goals and objectives, team leadership, team relationships, team roles and responsibilities, team communication, trust, and values. The study yielded two divergent definitions of team effectiveness, one reflecting the perceptions of the construction team and the other representing the viewpoint of the project owners. Azmy stated:

> Team effectiveness is defined as the team's capability to clearly establish, reach a consensus on, and comprehend the common project objectives, their respective roles and responsibilities for accomplishing designated tasks, and delivering a finished project of exceptional quality that aligns with the highest standards. Additionally, it involves the efficient adoption of owner's expectations throughout the project through the application of effective communication strategies. (p. 147)

The author noted that from these definitions, it could be inferred that team members emphasize an understanding of team effectiveness that centers on team goals and objectives, team communication, and team roles and responsibilities to ensure the project's alignment with owner expectations. In contrast, from the owner's viewpoint, team effectiveness was equated with the team's capacity to work harmoniously and accomplish the project within the stipulated timeframe and budget. This underscores the divergent perspectives on team effectiveness held by team members and project owners.

In this book, I offer a robust and all-encompassing maturity model specifically tailored for the effective management of construction teams (EMCTM2). This model, born out of a synthesis of existing literature about effective teams and insights gleaned from intensive interviews with nine seasoned project team members, serves as a comprehensive guide for enhancing the management of construction teams. As a result of the interviews, it has become evident that five foundational elements lie at the core of effective construction team management: organizational culture, team composition, managing the team life cycle, team building, and team leadership qualities and responsibilities. They serve as the cornerstone for achieving success and excellence team management in the construction industry. It is important to understand that these crucial elements exist on both a strategic level and an operational level. These five elements for effective management of construction teams may be modeled as below.

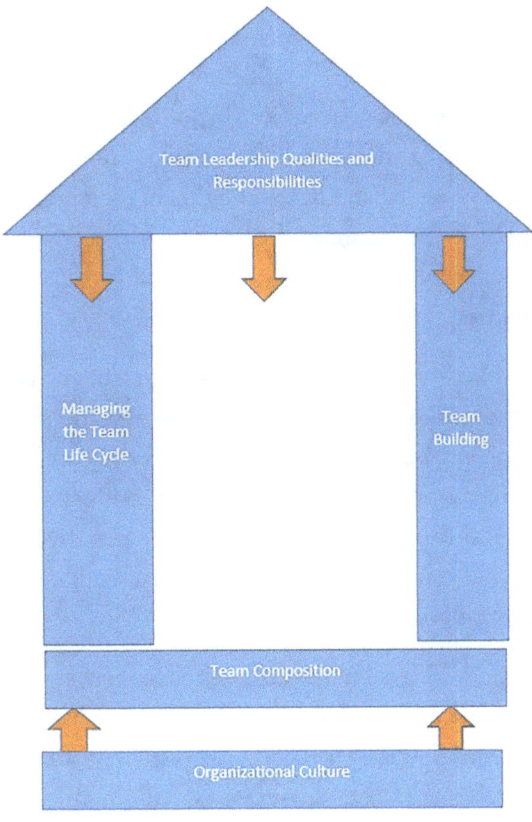

Figure 1
Effective Management of Construction Teams Model

Organizational Culture: This element underscores the significance of cultivating a positive and inclusive organizational culture to facilitate effective team management. It encompasses themes such as team composition, role clarity, trust-building, open communication of organization values with the team, recognition and rewards policy, empowerment, and avenues for learning and growth.

Team Composition: This element highlights the organization's pivotal role in selecting proficient team members, this element covers various subtopics, including defining selection criteria, promoting diversity, considering compensation, and continuously assessing and developing team members. It places a strong emphasis on the importance of clear objectives, diversity, qualifications, experience and expertise, and Interpersonal skills and capabilities.

Managing the Team Life Cycle: This element centers on strategies and actions for efficiently managing the project team life cycle, encompassing aspects such as effective communication, task planning and distribution, conflict management and dispute resolution, and team motivation. It underscores the importance of performance monitoring, goal clarity, trust-building, coaching, cooperation, and personal development.

Team Building: This element addresses a comprehensive range of strategies and actions for fostering effective team building within the construction industry. It covers areas such as role definition, skill enhancement, empowerment, communication, goal establishment, continuous evaluation, and ethical considerations.

It also delves into training programs, diversity promotion, resilience building, and team-building activities.

Team Leadership: This element offers an in-depth exploration of construction team leadership, including aspects like team dynamics, conflict resolution, communication, goal-setting, resource management, leadership experience, and nurturing interpersonal relationships. It highlights the importance of cultivating a positive work environment, adept problem-solving, empathy, delegation, accountability, and transparency.

By providing answers to pertinent research questions and developing a maturity model, I want to offer global benefits to the industry sector. The outcomes of this research can serve as a catalyst for organizations to improve their approach to team management, enabling them to modify and enhance processes, procedures, methods, and leadership skills pertinent to project team management. Moreover, the developed model holds promise in offering the industry and its firms a practical framework for human resource development activities in the future. Understanding the roles, skills, and career paths of project managers is crucial for selecting and developing effective managers capable of addressing challenges and delivering unique outcomes within constrained resources and timeframes.

Moreover, the absence of a standardized maturity model for evaluating the progress of team performance further complicates efforts to improve team management practices in the construction industry. Additionally, previous authors have not provided a maturity model that can serve as standardized criteria for the selection of team members, building the team, managing its development, and the competencies of the project leader to manage the construction team effectively. I offer here a maturity model for the effective management of construction teams (EMCTM2) tailored to address the complexities of managing construction teams effectively and fill the gap in construction team management literature and provide valuable and practical insights into enhancing the management of construction teams.

In the following chapters, I will dive into various aspects of construction team management to highlight the most effective strategies and techniques for enhancing performance. This will involve providing rigorous empirical insights and developing a practical model. I will examine the role of organizational culture, the composition of effective teams, strategies and actions for efficiently managing the team life cycle, and methods for effective team building.

Chapter 2

My Research Approach And Participants

After looking at previous work on effective team management, I decided to gather data from interviews (for depth) and a survey (for breadth). I embarked on this journey with six fundamental questions I designed to drive the study toward its predefined objectives. The development of this model represents a significant contribution to the body of knowledge and the theoretical underpinnings of construction project management, serving to bridge gaps in existing theories concerning team management within construction projects.

Crucially, the testing of the model after its development marked a pivotal step in the research methodology. Therefore, the inductive approach was deemed most suitable for this study, allowing me to tackle the research problem by generating a new theory. This approach supported my endeavor to comprehensively describe and analyze multiple facets of reality, constructing a profound understanding and capturing the intricacies of everyday human life and their perspectives.

All this meant I knew I wanted a mixed-methods approach combining face-to-face, semistructured interviews with nine expert project managers (qualitative technique) and a survey administered to 351 construction industry practitioners (quantitative technique). This research design addressed the absence of an information-concerned model for effective team management in construction projects. In order, my design methodology went through a literature review, formulation of interview questions, face-to-face interviews, data collection and analysis, model development, a questionnaire survey, and model validation. From there, I was able to mine deeply into the issue and arrive at valuable and practical recommendations regarding the effective management of construction teams from both leader and practitioner perspectives.

The primary objective of the in-depth interviews was to elicit responses that would address the research inquiries and facilitate the development of a maturity model. The interviews began with a concise explanation of the study's objectives, an overview of the participants' rights pertaining to confidentiality and ethical conduct, and a series of introductory questions beginning with collecting basic professional data about the interviewees, including their organization type, position within the organization, and years of professional experience. We then addressed organizational culture to gather data concerning the impact, role, and influence of organizations in the management of construction project teams. We turned then to team composition including team selection criteria, factors taken into consideration during the selection process, and the organizational role in the selection of team members.

We discussed in some depth managing the team life cycle so that experienced project managers could share insights on effective strategies for managing teams throughout their development, the needs of teams at different stages of development, and the most effective approaches for resolving disputes among team members. This part of the interviews explored the impact, strategies and techniques of the team life cycle that foster effective team management. We next addressed team building to concentrate on the actions, behaviors, and attitudes of project managers and organizations to manage construction teams effectively from their perspective. Finally, we talked about team leadership and style to see how project managers created conditions for their teams to achieve effectiveness, their approaches to handling issues and conflicts, their methods of supporting their teams and individual team members, and their actions to maintain team motivation and effectiveness that contribute to team management.

Semi-structured questions that drove the interviews included the following:

◊   What are the factors of an effective project team?

◊   What are the factors and techniques that influence the management of construction teams from the practitioners' perspective?

◊   Which strategies and techniques of organizational culture are essential for effectively managing construction teams?

◊   What strategies and criteria for team assembly are essential in steering and managing construction teams effectively?

◊   What strategies and techniques are indispensable for the effective management of the life cycle of construction teams?

◊   What strategies and techniques of team development/ building are essential for effectively managing construction teams?

◊   What are the fundamental leadership characteristics necessary for steering and managing construction teams effectively?

The nine project members, whom I will refer to as I1–I9, were actively engaged in the construction sector during the time of the interviews. These individuals represented client firms, consultancy firms, and contractor organizations. Each participant occupied distinct roles within their respective organizations, reflecting an extensive and nuanced background in the construction industry. Specially, the job titles covered were cost control and contract engineer, project manager, contracts manager, resident engineer, civil engineer, team leader, and construction engineer. Combined, they represented 200+ years of professional construction experience.

My next data collection came in the form a survey. The primary objective of the questionnaire survey was to validate the maturity model for the effective management of construction teams identified from the interviews with construction team practitioners. I emailed a link to the survey on a website to

500 well-respected project managers and their teams in the construction industry and received 385 responses, of which I was able to use 351. This distribution of the web address was intended to initiate a snowball effect, encouraging a network of construction industry practitioners to participate and contribute their valuable insights to the research effort.

To understand the data sources, I broke them down by several factors. First, I wanted to know the nature of the construction company the participants worked in. There were 35 responses (10%) from client companies, 150 (42.7%) from consultant companies, 136 (38.7%) from contractor companies, and 20 (8.5%) from supplier companies.

Next, I looked at their job descriptions. As there were quite a few, I'll skip listing them all and note that those with the same jobs totaling more than four were the following:

◊ Architectural Engineer-Electrical (21, 6%)

◊ Chief Surveyor (22, 6.3%)

◊ Construction Supervisor (23, 6.6%)

◊ Electrical Engineer (29, 8.3%)

◊ General Manager (23, 6.6%)

◊ Material Engineer (28, 8.0)

◊ Principal Structural Engineer (23, 6.6%)

◊ Quantity Surveyor (22, 6.3%)

◊   Resident Engineer (24, 6.8%)

◊   Site Engineer (23, 6.6%)

◊   Sr. Design Manager (23, 6.6%)

The participants had a considerable amount of experience with the majority of the participants having 6–10 years of experience at a total of 101 accounting for 28.8% of the sample size. Participants with 11–15 years of experience followed closely at 99 in total, a percentage of 28.2. These were followed by participants with 20+ years of experience at 57 in total, followed by those whose experience fell 16–20 years at 49, and lastly the least experienced with 1–5 years accounting for only 45 out of the 351 participants. This meant the data were collected from experienced individuals in the construction sector and would be best placed to support the investigation of effective management of construction teams and the development of a maturity model for effective management of construction teams.

From all these participants, I next coded and sorted the data to extract commonalities. This led to finding my indices, or items of measurement, which I was then able to categorize under five specific variables for the maturity model, as follows:

◊   Organizational Culture

◊   Team formation and composition strategy are essential factors for construction team management.

◊   Communication with the project team is an essential factor for construction team management.

◊ Employee development is an essential factor for construction team management.

◊ Team members' support is an essential factor for construction team management.

◊ Fostering a culture of collaboration, trust, and respect is essential for construction team management.

◊ Recognition and reward policies are essential for construction team management.

◊ Team Composition

◊ Structured selection criteria and processes are essential for construction team management.

◊ Qualifications, experience, and expertise are essential for construction team management.

◊ Interpersonal skills and capabilities are essential for construction team management.

◊ Team Development/building

◊ Empowerment, delegation, and team dynamics are essential for construction team management.

◊ Building effective teamwork and a collaborative environment is essential for construction team management.

◊ Personal growth and character development are essential factors for construction team management.

◊ Managing the Team Lifecycle

◊ Communication is essential for construction team management.

◊ Team environment and structure are essential for construction team management.

◊ Goal setting with clear roles and expectations is essential for construction team management.

◊ Planning and task distribution are essential for construction team management.

◊ Team Leadership Qualities

◊ Recognition and appreciation are essential leadership factors for construction team management.

◊ Goal setting and communication are essential leadership factors for construction team management.

◊ Continued improvement and problem-solving are essential leadership factors for construction team management.

◊ Team motivation and well-being are essential leadership factors for construction team management.

◊ Collaboration and trust are essential leadership factors for construction team management.

If you've checked the Table of Contents, you can see this roughly outlined the topics of my chapters, starting with Chapter 4.

Before discussing the details, outcomes, and recommendations regarding these theme variables and shared indices, as much as I'd like for you to just take my word for it that they are all correct, I need to show how the data were tested for various factors of validity and then pivot to test the validity of the model itself.

Chapter 3

Evaluating My Data And Model

Analysis of the interviews focused on shared themes by coding the initial data, extracting pertinent keywords and phrases, and amalgamating similar categories and concepts. This method facilitated a more comprehensive understanding of the data, identifying patterns and connections among various elements. The analytical journey culminated in the final stage, where I took these themes and merged into an overarching concept. I reviewed the data, followed by editing, coding, checking for inconsistencies and duplications, and subsequently entered the information into the Statistical Package for Social Sciences V.29 software. This process formation laid the foundation for the development of the new effective management of the construction team maturity model (EMCTM2).

In addition to the information from the nine interviews, I sent out 500 random emails with a link to an onsite questionnaire to construction workers through an electronic survey platform. I initially collected a total of 385 responses for a response rate of 77%, which is in line with recommendations from Fincham (2008) and many others on an acceptable response rate for questionnaire survey research set at a threshold of 60%. Out of these responses received, I found that 26 had incomplete information, so expunged them to retain a total of 359 complete responses.

I then checked for outliers with the Kolmogorov-Smirnoff and Shapiro-Wilk tests of outliers. It was clear the data came from a normal distribution.

Table 1

Kolmogorov–Smirnoff and Shapiro–Wilk Tests of Outliers

	Kolmogorov-Smirnov[a]			Shapiro-Wilk		
	Statistic df Sig.			Statistic df		Sig.
OC TC	.196 359 < .001			.861 359		< .001
	.210	359	< .001	.910	359	< .001
MTLC	.275	359	< .001	.816	359	< .001
TD TL	.333	359	< .001	.814	359	< .001
	.184	359	< .001	.874	359	< .001

OC = Organizational Culture; TV = Team Composition; MTLA = Managing the Team Life Cycle; TD = Team Development; TL = Team Leadership

The assumption of the Kolmogorov-Smirnov and Shapiro-Wilk's tests are that a dataset comes from a normal distribution with a confidence level of 95% (Ryu, 2011). If the results of the tests result in a p-value of less than or equal to .05, then the null hypothesis or primary assumption is accepted. If the 95% confidence level is broken and the p-value returned is above .05, then the null hypothesis or primary assumption is rejected and, it is determined therefore that the data is not normally distributed. In the case of the current study, the significance level for all the constructs is .000 which means that the data comes from a normal distribution or is normally distributed. I found that the dataset did not have any significant outlying aspects, as reflected by a stable range of values and a significant value of 0.000 at a 95% confidence level.

I also checked for normality, for which I used skewness, the distance of the data values from the normal distribution based on the symmetry of the data, and the Kurtosis test, which determines the level of tailing of values in the dataset when considering the normal distribution of the data.

Table 2

Skewness and Kurtosis Test

	N	Mean	Skewness		Shap-iro-Wilk	Kurtosis	
	Statistic	Statistic	Statistic	Std. Error	Statistic df	Statistic	Std. Error
OC TC	359	4.2129	1.082 .129			1.016 .257	
	359	4.3092	.033		.129	.042	.257
MTLC	359	4.1532	1.305		.129	.927	.257
TD	359	4.0325	1.062		.129	1.259	.257
TL	359	4.2006	.729		.129	3.620	.257
Valid N (list wise)	359						

The skewness and kurtosis values confirmed that the data were normally distributed. This is justifiable by looking at the skewness values of all of the constructs in the dataset, which fall squarely between -3 and 3; on the other hand, the kurtosis values of the data values fall between -10 and 10 (Brown, 2016; Hair et al., 2020). Skewness and kurtosis values for normally distributed data should fall between -3 and 3 and -10 and 10, respectively. With a dataset satisfying these values, it is possible to proceed to other aspects of analysis because the data satisfies the general requirements for normality.

With univariate outliers confirmed as missing from the dataset, I needed to examine the data for any presence of multivariate outliers. I used the Mahalanobis (1927) distance test, which determines how far responses making up a given construct are from the normal distribution of the dataset's constructs' central distribution. I was hoping to find all the Mahalanobis distance values of the cases in the dataset under all the constructs that have been included in the test when compared to the cumulative distribution of Chi square with the same degrees of freedom, which are higher than 0.001.

Again, for those who like tables:

Table 3

Mahalanobis Distance Test for Multivariate Outliers

Case number	Mahalanobis d-squared	ChiSq p1
289	28.6114	.00003
317	25.1107	.00013
319	29.9245	.00002
321	37.8269	.00000
328	27.9875	.00004
331	21.6357	.00061
336	24.8347	.00015
358	23.7372	.00024

Any value that is lower than this is considered to be a multivariate outlier, which should be eliminated from the dataset to ensure the dataset retains its validity. Each of these case numbers, a total of eight, was deleted from the dataset, bringing down the number of cases in the dataset from 359, to 351 going forward.

Considering the construction of the variables of the current study, the assessment of the measurement model could only be undertaken at the formative level. This is unlike in cases where variables of a study have subvariables that make up each of the variables, which then calls for the assessment of the first and second order variables. My only available variables were the first order and were assessed as such to determine critical elements of reliability in line with prescribed approaches for the same. Essentially, in the assessment of the measurement model, I was interested in examining the reliability and validity with specific consideration to convergent and discriminant validity in the measurement model.

Construct validity is synonymous with the internal consistency of the constructs of measurement commonly assessed using the Cronbach alpha coefficient at a threshold of 0.7. Similarly, convergent validity, which is the measure of the relativity of constructs, is pegged at the threshold of 0.7 for each indicator's loading with loading equal to or lower than 0.4 being identified as problematic as far as convergent validity is concerned and recommended for forthwith removal.

Under the theme of organizational culture, "empowerment and autonomy are essential for construction team management" had an indicator loading of 0.052. Under managing the team life cycle, "Conflict management and dispute resolution are essential for construction team management" had an indicator loading of 0.47. Under team leadership essential qualities of "emotional intelligence and team dynamics" had an indicator loading of 0.343, "team inclusivity and diversity" had an indicator loading of 0.028, "leadership guidance" had an indicator loading of 0.233, and "empowerment and ownership" had an indicator loading of -0.042.

I chose these items of measurement redaction from the data instrument to allow for increased validity and reliability as measured by parameters such as composite reliability and average variance extracted. Each of the remaining measurement items for the formative variables of the study assessed recorded indicator loadings above the 0.5 threshold and the construct reliability metrics of Cronbach alpha, save for the variables of team composition and team development, and composite reliability, and all recorded impressive indices above the acceptable thresholds of 0.7. The items of measurement making up each construct scored high indicator loadings, which then contributed to increased scores in Cronbach alpha internal consistency coefficient, composite reliability, average variance extracted (AVE) and the variance inflation factor (VIF) statistics, which are all critical in assessing validity and reliability of datasets. Noteworthy is that the variable of team composition barely satisfied the AVE and Cronbach alpha thresholds, as well as the variable of team development, which also barely satisfied the Cronbach alpha coefficient. This may be as a result of very few measurement items allocated to the variables when compared to the other variables in the same model. I noted this for referral as a limitation of my study.

The following table shows all of this together with statistics for the AVEs and full collinearity of the formative variables of the study, confirming that they (save for one instance already discussed) all fall within acceptable thresholds of above 0.5 and below 10, confirming the positive convergent validity of the measurement model.

Table 4

AVEs and Full Collinearity of the Formative Variables

Measurement Item Essential for Construction Team Mgt.	Item Loading	Composite Reliability	Cronbach's alpha	AVE	VIF (Full Col-linear-ity)
Organizational Culture		0.857	0.797	0.506	3.058
Team formation and composition strategy	0.726				
Communication with the project team	0.52				
Employee develop-ment	0.784				
Team members' support	0.818				
Culture of collab-oration, trust, and respect	0.795				
Recognition and reward policies	0.57				
Team Composition		0.749	0.497	0.498	1.258
Structured selection criteria and processes	0.718				
Qualifications, expe-rience, and expertise	0.688				

Interpersonal skills and capabilities	0.711				
Managing the Team Life Cycle		0.855	0.795		0.501
Communication	0.604				
Team environment and structure	0.619				
Goal setting, clear roles, and expectations	0.786				
Planning and task distribution	0.843				
Team performance management	0.765				
Coaching and skills enhancement	0.586				
Team Development		0.808	0.643	0.585	1.663
Empowerment, delegation, and team dynamics	0.743				
Team Leadership Qualities		0.889	0.842	0.62	3.841
Recognition and appreciation	0.747				
Goal setting and communication	0.862				
Continues improvement and problem-solving	0.884				
Team motivation and well-being	0.633				
Collaboration and trust	0.785				

Next, I turned to the discriminant validity, which measures the extent to which each construct is distinct, including the overlap, from the other in a dataset. This measurement is based on the square root of the AVE, which should be higher than all the other correlations between the construct and the other constructs in the model (Chin, 2010; Hair et al., 2020; Santos & Cirillo, 2021). Having removed the items of measurement that affected the other indices of validity and reliability, I found that the discriminant validity based on the squared average variance extracted was attained. The table below shows that the model passed the acceptable discriminant validity measure with all the square root average variance extracted for all constructs being the highest in comparison to their correlations with other constructs in the model. That is, for organizational culture, the square root average variance was 0.712, for team composition, it was 0.706, for managing the team life cycle, it was 0.78, for team development, it was 0.765, and for team leadership, it was 0.787.

Table 5

Square Root Average Variance

	OC	TC	MTCL	TD	TL
OC	0.712				
TC	0.425	0.706			
MTCL	0.649	0.421	0.708		
TD	0.562	0.271	0.522	0.765	
TL	0.685	0.398	0.695	0.619	0.787

With the research through standard analytical approaches' having confirmed the goodness of fit of the model, it was now tenable to use the model for the conduct of structural assessment. In other words, I could then review the strengths of the different relationships of variables in the model. This involved the checking and reporting of the significance of the path coefficients recorded and the R^2 statistical coefficient for endogenous variables in the model. As captured in the conceptual model, a causal link connected the different variables in the study with the assumption being that these variables and their connection in the model represent the maturity model for the effective management of a construction team. Using SPSS AMOS v.29, I calculated the standardized regression coefficients of the relationships together with their corresponding significance values (p-values). I also report the R^2 coefficient as it communicates the percentage of the variance of the observed variable exerted by the influencing variable. Sarstedt et al. (2021) defined the regression path coefficients as the estimates of the functional relationship between variables, which therefore defines the level of dependence of one variable to the other, and further defined the R^2 coefficient as the proportion of variance in the dependent variable predicted by the independent variable. I've put the results in the form of both a table and a figure.

Table 6

The R² Coefficients

Path	Path Coefficient	p-value	Statistically Significant?
OC->TC	0.340	< 0.001	Yes
TC->MTLC	0.125	0.009	Yes
TC->TD	0.022	0.341	No
TL->TC	0.129	0.007	Yes
TL->MTLC	0.727	< 0.001	Yes
TL->TB	0.643	< 0.001	Yes

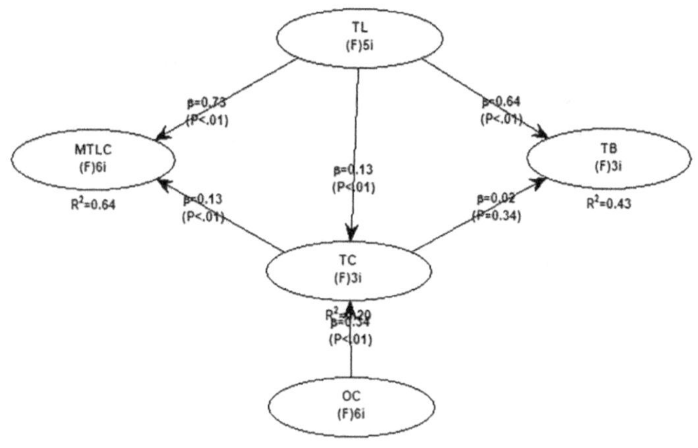

Figure 2

The R² Coefficients

I tested a number of relationships between the variables. I tested the relationship between organizational culture and team composition and found that organizational culture had a positive effect on team composition at 0.340 at a significant p-value of < 0.001. This implies that when organizational increases by unit one, it leads to the improvement and/or increase of the team composition by 0.340. The R^2 value recorded 0.151, implying that organizational culture had an impact on change of team composition of up to 15.1%.

I looked at the relationship of team composition and effective management of the team life cycle, and the results of the path analysis revealed that team composition positively impacted the management of the team life cycle at a regression coefficient of 0.125 at a significant p-value of 0.009. This implies that when team composition goes up by unit 1, the management of the team life cycle improves by 0.125. The R^2 squared coefficient recorded was 0.066 to indicate that although the relationship was positive, it was not supported because of a low size of effect. According to Ozili (2023), the R^2 value should fall between 0.1 and 0.9 for the size of the effect to be acceptable.

I looked at the relationship between team composition and team development and found team composition had a mildly positive effect on team development at 0.022, but this relationship was statistically insignificant with quite a high p-value of 0.341, indicating that it broke the confidence level of 95% and therefore rendered the result statistically insignificant.

The fourth relationship I examined was between team leadership and the management of team life cycle, which recorded a regression coefficient of 0.727 at a significant p-value of < 0.001. This was the strongest regression coefficient, and it indicated that with the increase of team leadership by unit one, the management of the team life cycle improved by 0.727. The effect size of this relationship measured by the R^2 coefficient recorded 0.048, which nullified the significance of the effect as it did not meet the threshold. The case was the same for the relationships between TL and TC as well as TD which, although accepted and statistically significant regression coefficient, had R^2 values that did not exactly meet the required threshold to validate an effect size.

I was thus confident to use the data as the basis for the remaining parts of the study. To do this, presentation of the characteristics of the data using descriptive statistics lets us see what the sample represents in terms of different parameters. I first assessed the measurement model and then the structural model, which is also the path analysis.

In the initial step of measurement model assessment, the reliability and validity of the dataset are examined using a number of tests. This examination of reliability and validity is based on the relationship between constructs and the items of measurement used to measure the construct. On the other part of the structural model assessment, the goal is to test the relationships between constructs in the model and through this, determine the acceptance or rejection of the hypothesis of the research if applicable, or make deductions of the strength of the relationships between the constructs thereof. Conceptual frameworks are commonly made up of reflective or formative variables. Understanding the type of variable in the conceptual model is essential for effectively assessing the measurement model. A reflective variable in a conceptual model is one that is affected by the latent variable and is thus interchangeable with other reflective variables of the latent variable without directly affecting the construct of the latent variable. On the other hand, a formative variable is one that defines the latent variable and is thus not possibly interchangeable without considerably altering the construct of the latent variable it forms.

So, there were a total of five variables (organizational culture, team composition, team development/building, managing the team lifecycle, and team leadership qualities). As there were no underlying sub-constructs that reflected the constructs, the constructs were the main component of their own and only supported by the measurement items, which made them formative variables by default considering the description of formative and reflective variables.

As I said in Chapter 2, the primary goal of the survey was to validate the maturity model created from the interviews. The qualitative analysis of the data collected from 351 experienced individuals in the construction sector refined and validated the model. The assessment of the measurement model for the current study revealed that a number of measurement items, though some had lower indicator loadings than I expected and had thus to be expunged from the model. These removed items affected the other indices of validity and reliability, so they were eliminated from the constructed model. The final model is elucidated below in a table and figure.

Table 7

Effective Team Management Model

Effective Team Management Model Dimensions	Factors and Techniques Essential for Construction Team Management
Organizational Culture	Team formation and composition strategy Communication with the project team Employee development Team members' support Foster the culture of collaboration, trust, and respect Recognition and reward policies
Team Composition	Structured selection criteria and processes Qualifications, experience, and expertise Interpersonal skills and capabilities
Managing the Team Life Cycle	Communication Team environment and structure Goal setting, clear roles, and expectations Planning and task distribution Team Performance Management Coaching and Skills Enhancement
Team Development/ building	Empowerment, delegation, and team dynamics Building effective teamwork and a collaborative environment Personal growth and character development
Team Leadership	Recognition and appreciation Goal setting and communication Continues improvement and problem-solving Team motivation and well-being Collaboration and trust

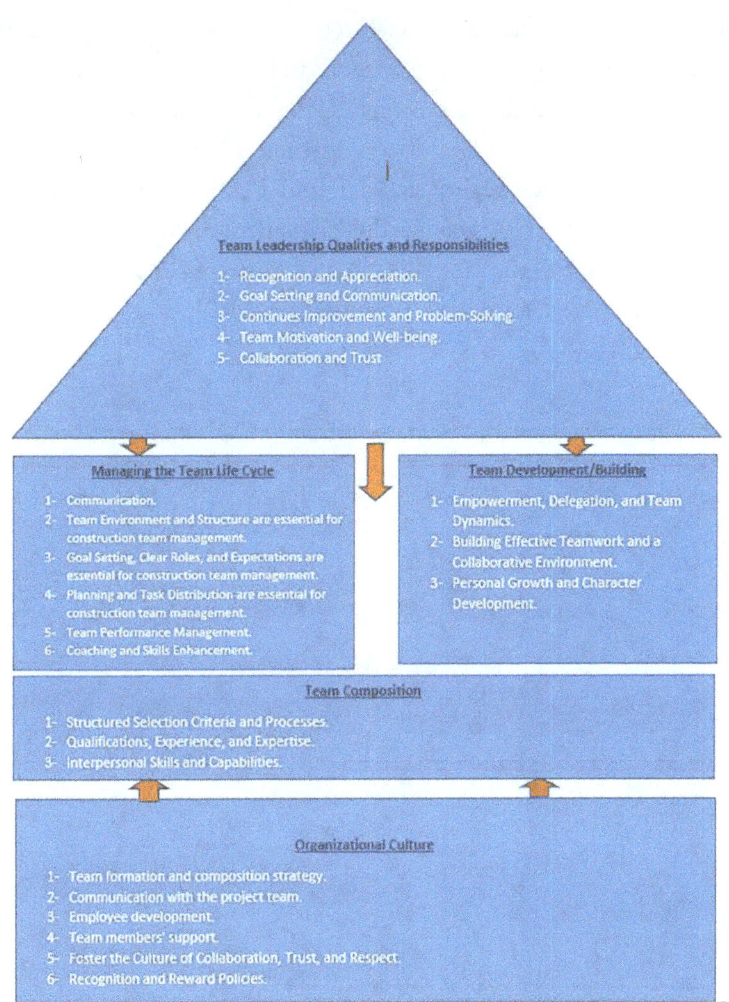

Figure 3

Effective Team Management Model

Before going further in my research, I disseminated this model to many practicing project managers and team members actively engaged in construction projects for their valuable feedback and validation. This iterative process further enhanced the model's reliability and practicality, and I found I did not need to alter it substantively. This was particularly encouraging considering Merriam's (1998) warning that "In qualitative research, the reality is characterized as holistic, multidimensional, and constantly evolving" (p. 202).

Chapter 4

Organizational Culture

The project team operates within a distinct social and cultural framework embedded within the organization. This cultural milieu significantly influences various aspects of team dynamics, including development, growth, synergy, performance, and long-term viability. Within this cultural context, organizational elements such as structure, motivation systems, information flow, knowledge exchange, employees' perceptions, responsibilities, problem-solving mechanisms, support for career development, and rewards and recognition systems all play a pivotal role in shaping the effectiveness of team management, either through positive or negative contributions.

"Organizational culture encompasses an organization's collective expectations, experiences, philosophy, and values that bind it together, manifesting in its self-image, internal dynamics, external interactions, and future outlook" (Fokina & Krupnov, 2020, p. 1). Furthermore:

Organizational culture becomes deeply ingrained in employee behavior, spreading and permeating unconsciously. It acts as a social adhesive, fostering a sense of belonging within the organizational milieu. Employees are naturally inclined to internalize the organization's culture because it satisfies their social identity needs and imbues their experiences with meaning. Organizational culture plays a pivotal role in the process of comprehension, aiding employees in grasping what to perpetuate and why certain occurrences transpire within the company. It also facilitates employee understanding of their expected roles and interactions with trusted colleagues (Fitria, 2018, pp. 83–84).

The collective values and norms shared among employees in an organization are indicative of its commitment to teamwork and its effectiveness in team management. Additionally, the organization's work policies and its overall environment exert a significant influence on a team's stability and productivity, as highlighted by Piwowar-Sulej (2021), who added, "Organizational culture, much like the subconscious, wields an impact on employees' aspirations, attitudes, and behavior." Moreover, Katane and Dube (2017) underscored the pivotal roles played by the organizational culture that included good leadership, trust, communication, and team commitment in the success of project teams. Mohsen et al. (2020) noted that elements of organizational culture have a positive influence on employee performance, particularly in achieving goals, fostering coordinated teamwork, emphasizing customer orientation, and bolstering organizational culture strength. A well-functioning organizational culture serves as a motivational catalyst for employees, driving higher- quality performance through collaborative and ethical work behavior (Lee 2020). Hence, "Effective organizational culture entails shared values and a common mission to cultivate a sense of teamwork" (Pathiranage et al., 2020, p. 528). Bresman and Zellmer-Bruhn (2012) investigated the interaction effect between team and organizational structure and found that

> organizational structure supports external team learning under conditions of less team structure. Specifically, when teams have less team structure, the relationship between organizational structure and external team learning is positive. This structure substitutability finding suggests that although more organizational structure, on average, hurts external team learning, there are situations in which it helps.

Employees' collective values and norms indicate their commitment to teamwork and effectiveness in team management. The organization's work policies and overall environment influence a team's stability and productivity. These make up the organizational culture. In the context of globalization, new approaches to work organization, especially in managing innovative activities, significantly impact employees' mindsets, altering their attitudes toward team-oriented values and affecting their perceptions and assessments of innovation's importance for the enterprise.

Putra et al. (2020) have reminded us, however, that organizations must furnish a conducive work environment that includes both physical elements such as office layout, cleanliness, air quality, color schemes, and lighting. Organizational responsibilities include providing necessary resources, establishing a safe workplace, and offering support to teams when required. Salas et al (2018) emphasized that "it is critical to have teamwork supportive organizational conditions and environments where psychological safety can flourish and be a mechanism to resolve conflicts, ensure safety, mitigate errors, learn, and improve performance" (p. 1). Keck and Tushman (2017) highlighted that

> the longer the period of stability in a team environment, the less change in members, the greater the mean tenure, and the greater the homogeneity. Whereas periods of equilibrium are associated with low change and high homogeneity, organizations that survive dramatic environmental shifts have heterogeneous executive teams that display both stability and the capacity for change.

Furthermore, effective management involves guaranteeing that projects are equipped with a sufficient number of skilled laborers.

My interviews with construction industry experts echoed many of these conclusions regarding organizational culture and addressed the influence, role, and criteria of organizational culture in shaping the nurturing and effective management of construction project teams. Several talked in particular about communication with the project team, I1 stressed communicating "openly and transparently," I3 advocated regular reporting on project status, and I9 said team member should be allowed to speak up about any issue in the project openly. I2, I6, and I8 talked about encouraging communication, with I8 noting tools and resources need to be provided to facilitate communication. I5 suggested arranging regular meeting for problem-solving, and I2 discussed promoting communication both among project members and between top management and the project team.

Regarding employee development, I3, I4, I8, and I9 specifically mentioned the need for training team members; I5 suggested organizing regular workshops, seminars, and related courses; and I3 said leaders need clear strategies and policies for employees' career path development. This led to discussion of supporting team members to build, as I1 put it, "team cohesion." Specific techniques included correcting and providing feedback (I2), fair treatment (I9), giving employees "space" (I6), providing a safe and healthy environment, (I5), and being "close" to the team (I4). I3 said top management should arrange essential tools such as cars and computers, offer legal advice as needed, and provide updated rules, standards, and specifications. I8 encouraged coaching and mentorship either through in-house programs or by partnering with external coaches and mentors. I8 called for providing a "culture of safety."

Almost everyone discussed the need to empower members and provide them with autonomy where possible. I1, I4, I6, I8, and I9 all talked about the importance of delegating decision-making, though I2 talked about the need to monitor progress with regular reporting, which I9 discussed as providing clear performance evaluation criteria. I6 said leaders need to clearly define roles and responsibilities for every team member and to keep the team informed about the organization's vision and values. I8 emphasized flexibility such as remote work options, flexible hours, and job sharing.

Six of the nine interviewees discussed the need to foster collaboration and trust and to respect people's cultures. I7 and I8 talked about encouraging feedback. I9 said organization management should meet their teams and celebrate lunch with them to encourage collaboration between team members, and I5 suggested sharing regular sports, picnics, and entertainment activities with the team. I8 said one should "foster a culture of collaboration." Such approaches also united teams around organizational strategy. I8 added, "Foster diversity and inclusion by implementing policies and practices that support diversity, equity, and inclusion." I6 said it was all about "Filling the team with the right people that having problem-solving, decision-making, and interpersonal skills." Finally, three interviewees called on the need for a strong recognition and rewards policy, with I1 stressing one should reward teams for taking risks (one assumes successfully), and I2 liked awarding high-performance members.

Generally, the interviewees were positive that a good team environment where there is collaboration between team members, effective communication, and fluid access to information and expertise from member to member, helps in ensuring that the team remains cohesive and committed to delivering the best outcomes from the projects in their hand. Some stressed that a clear structure of the team was of great importance, though this is tempered by the belief that a friendly, less rigid environment whereby each team member can freely interact even with the team leader came forth as the best approach to ensuring team success. From a practical perspective, in teams where members interact much during the process of executing their duties, they are likely to share more, collaborate more, be innovative, and essentially realize more problem solving and success.

A large part of organizational culture is thus clearly team management. Dolgov et al. (2006) have emphasized the pivotal role of organizational culture and delineated factors that influence the implementation and development of team management in organizations. These factors include the following:

◊ Organization leader's commitment to team development;

◊ Ability of management team members and top managers to lead their functional teams;

◊ Clarity of the organization's strategic and operational goals, which should be well-known to all employees;

◊ Existence of effective feedback mechanisms between management levels;

◊ A transparent system of material and non-material incentives, aligning employees with individual and collective effectiveness; and

◊ Efficacy of organizational philosophy as a management tool.

Fong and Kwok (2009) looked at organizational culture and knowledge success within contracting firms at both project and organizational levels and found that "Clan culture is prevalent at both project and organization levels, signifying that the culture within contracting firms heavily relies on honest communication, respect for individuals, trust, and strong interpersonal relationships" (p. 1353) The authors argued that the construction industry in particular, being both project-based and people-centric, underscores the importance of these cultural aspects.

The interviewees acknowledged that each strategy and technique of organizational culture, including but not limited to proper alignment and establishment of teams so that the best-suited individuals can work on a project and find the best fit among colleagues, was of importance to the organization. The role of the rewards system and mechanism toward building an effective organizational environment that can support the development of effective teams was also prominent from primary research in agreement with extant literature. For instance, I1, I2, and I5 indicated that wages should be reasonably designed and employees in the project teams should be rewarded for over performing, taking risks, and resolving problems in an effort to strengthen and foster a stronger team relationship.

Interviewees also emphasized the importance of an organizational culture that prioritizes respect, trust, and collaboration as foundational for effective team management. Ensuring that team members are aware of the organization's culture, vision, mission, and values was considered essential for reinforcing alignment with organizational goals and creating a shared sense of purpose among team members. Encouraging team members to express and share their ideas was recognized as a valuable technique for promoting collaboration and innovation, fostering an inclusive environment where diverse perspectives are considered, leading to more effective decision-making. Moreover, recognizing and rewarding high-performance achievements were seen as motivating and reinforcing positive behavior, encouraging team members to strive for excellence, and fostering a sense of accomplishment. Providing competitive and fair wages for team members was viewed as integral to a recognition and rewards policy, acknowledging the value of team members' contributions and enhancing their job satisfaction.

The interviews underscore the significance of organizational cultural factors and techniques influencing effective team management. Each factor and technique plays a crucial role in creating a collaborative, inclusive, and empowered team environment that fosters trust, respect, and recognition. The research interviews not only revealed these key themes but also emphasized the organization's dedication to fostering a culture that values effective communication, skillful team composition, collaborative problem-solving, career development, and ongoing learning. These cultural factors collectively contributed to the successful management of construction teams and the achievement of organizational goals.

Therefore, the research in considering the practical aspects of the primary research findings as well as literature related notes that, in a usual construction environment, the success is indeed set by the culture that the organization is working with. For instance, if an organization operates from a culture of collaboration and information sharing, it is very likely that competing interests in the organization will be limited and employees in the teams will be less averse to sharing ideas and ensuring that all employees work toward achieving the goals of the organization. If the team is composed of people who fit together well and are competent in their jobs, then the execution will be top notch and this is a function of the organizational culture under team composition. It is clear that the establishment of lines such as lines of autonomy and clear guidelines for the employee rewards policies that are fair and encourage the employees to improve, it is certain that the employees are likely to perform better and with this, there will be more success of the project team. The fact that the organizational culture variable in the validated model was found to have a positive impact with the other variables in the model is a testament to the level of importance of organizational culture in the success of construction project teams.

Thus, organizations prioritizing these aspects can expect a more productive, collaborative, and motivated workforce, ultimately leading to project success. Therefore, organizations should cultivate and communicate a culture that values respect, trust, and collaboration, serving as the foundation for cohesive teamwork, alignment with organizational goals, and a shared purpose among team members. Organizations must adopt a deliberate approach to assembling teams with the appropriate skill sets to enhance overall team performance, facilitating better problem-solving and decision-making to achieve project goals and deadlines. They should actively provide opportunities for skill development and career advancement, as well as encourage delegation to benefit both individual team members and the organization, thereby ensuring a skilled and motivated workforce and fostering responsibility among team members. Establishing transparent evaluation processes that clarify performance expectations and align them with team success is vital for maintaining team morale and motivation. Implementing recognition and reward policies to encourage team members to excel fosters a sense of accomplishment while providing competitive and fair wages acknowledges the value of team members' contributions and enhances job satisfaction.

Chapter 5

Team Composition

The formation of teams and their composition have been subjects of extensive study in such social science fields as human resource management, sociology, and psychology. Team formation significantly influences teamwork, directly shapes team performance, and can deeply affect team members' attitudes and behaviors, particularly when the right mix of individuals with complementary skill sets is achieved.

It's generally understood that poor teams are plagued by such issues as improper or inadequate skill sets, interpersonal conflict, lack of team spirit, and simple incompetency. But what makes for a good team, and how can their formation be facilitated?

The significance of organizational culture extends beyond understanding the operational parameters within which individuals and teams function; it also plays an essential role in individuals' selection criteria and processes. The initial step in assembling cohesive teams is to cultivate a comprehensive understanding of the organizational context in which these teams will operate. Consequently, the team member selection process should prioritize team building and collaboration. This implies the need for an appropriate mix of team members, established norms, and clearly defined tasks to foster positive team spirit and dynamics.

The literature within the construction industry has shed light on various studies and models of team composition, which have predominantly centered on the skills, knowledge, and behavior of team members. Rahmani et al. (2021, p. 17) conducted an investigation into the knowledge, skills, attributes, and experience (KSAE) required by individuals within the context of relational contracting in construction projects. Through content analysis of interviews, the authors identified sixteen behavioral factors and four knowledge and experience areas, which were further analyzed in alignment with Belbin's (1981) team role clusters framework. The results indicated that while all three role clusters (action-oriented, people-oriented, and thought-oriented) are valuable, in relational contract arrangements, team members are particularly required to be proficient in people-oriented roles more than in the other two role clusters. Trust, mutual dependency, freedom of expression, established relationships, and open access to participants are pivotal indicators of a construction team's maturity. There are multiple preferred team roles. Notably, the most preferred role of quantity surveyors, clients, and contractors was the resource investigator, whereas architects, civil engineers, and service engineers favored roles such as planner, implementer, and specialist, respectively. The authors pointed out that assessing the preferred team roles of all potential team members in a construction project before their inclusion is essential.

If done correctly, this will help to harmonize team members' personal characteristics with the team dynamics and other factors that impact teamwork. Combining personal and technical skills is crucial for optimizing team performance. Integrating personality-based skills can encourage positive employee behavior and contribute to overall project success. Mathieu et al. (2017) reviewed the evolution of team research and highlighted a shift from focusing on individual team members or individual-versus-team comparisons to centering on the team itself and larger systems of teams. The authors emphasized that team composition can be viewed in terms of two overarching themes: which characteristics should be considered, and how these characteristics are distributed within the team. Bell and Brown (2015, p. 181) argued that effective team staffing necessitates consideration of team members' knowledge, skills, abilities, and other characteristics (KSAOs), as well as the configuration of team member KSAOs and their interrelationships. They stress that proper team staffing encompasses individual-level factors such as a team member's ability to execute technical aspects of their role, team-level factors like the complementarity of KSAOs within the team, and organizational factors concerning how the team contributes to organizational success.

Bell et al. (2018) looked at team composition and teamwork to identify four foundational concepts for describing the relationships between team composition and teamwork: team member attributes, operationalization at the team level (e.g., diversity, team means), the context, and temporal considerations (time). They further emphasized that each team role requires specific characteristics suitable for its purpose and responsibilities within the team (Zainal et al., 2020, p. 556). Thus, it is important to acknowledge that team members' personalities play a pivotal role in effective team composition, and neglecting personality variables during team role assignments may impact the overall success of the team (Gilal et al., 2017).

Moreover, Payne and Harper (2020) argued that the assessment of personality traits can be employed to understand an individual's capabilities and consequently place them in roles where they can function effectively within a team. The author focused on the "big five factors" (BFF) for personality traits and explored how they might influence the selection process for project team members. Specifically, the authors considered personality traits related to team performance, team building, team efficiency, teamwork, job performance, and collaboration within construction project teams. Payne and Harper found that

> conscientiousness has the highest correlation mean and neuroticism has the lowest correlation mean for team performance, while extraversion and agreeableness do not share a common correlation mean, making these traits heterogeneous. Conscientiousness, neuroticism, and openness to experience share a common correlation mean, making these traits homogeneous. Overall, the results of the study showed that personality traits do influence team performance. (p. 1)

My interviews with construction industry experts echoed many of these ideas and offered new and/or more detailed approaches to crucial aspects of team composition that contribute to the effective management of construction teams. These factors comprise selection criteria, factors deliberated upon during the selection process, and the pivotal role played by the organization in the selection of team members.

Regarding interpersonal skills and capabilities, I2 talked about the desirability of a passion for learning and personal development. I4 advised looking for proactive members, and I6 said leaders should "Select a member who can bring a diverse range of viewpoints and ideas to the table." Almost all discussed the need to engage strong communicators with good people skills as well as selecting people based on the compatibility of their personal skills with the duties of the position. I1 advised prioritizing skills and knowledge over certifications, and I3 stressed the desirability of "soft skills."

In discussions of considering members' qualifications, experience, and expertise, while some interviewees called for checking members' formal education, almost all stressed examining the members' experience, particularly with similar work. Regarding the actual structured selection criteria and process for assembling teams, most stressed the need for clear and well-defined job requirements, which then should be matched with the specific skills and background of the employees. I1 talked about the resulting need for "diversity in the team members," and I8 said, "During the interview, ask questions that assess their skills, experience, and fit with the organization's culture." I9 added that

"The team leader's feedback should be considered in the team selection process."

From the interviews, as supported by the survey, three major factors emerged, each significantly contributing to the effectiveness of the construction team and facilitating efficient management practices, and each of these were supported by multiple criteria, as follows:

◊ Structured Selection Criteria and Processes

  ◊ Establishing a specific selection process: A clearly defined process tailored to the project's needs ensures a systematic and fair selection of team members.

  ◊ Alignment with project requirements: Analyzing project needs and selecting individuals accordingly ensures a match of skills and expertise to project demands.

  ◊ Defining clear team goals and roles: Clarity in role definition and team goal setting ensures the selection of suitable individuals for specific tasks.

  ◊ Objective evaluation and continuous assessment: Objective assessment of candidates and ongoing evaluation of selected members help maintain team efficiency and effectiveness.

  ◊ Development and feedback: Investing in the development of selected members and considering input from team leaders enhances team adaptability and performance.

◊ Diversity considerations: Recognizing diversity in backgrounds, experiences, and perspectives enriches the team's problem-solving capacity and creativity.

◊ Qualifications, Experience, and Expertise

◊ Educational qualifications and project experience: A degree certificate and previous experience in similar projects ensure a foundational knowledge base and practical understanding for effective contributions.

◊ Administration and team management skills: Administrative competencies and knowledge of project team management are crucial for streamlining project-related activities.

◊ Interpersonal Skills and Capabilities

◊ Communication and interpersonal relationships: Strong communication skills and the ability to maintain good relationships within the team are pivotal for effective collaboration and conflict resolution.

◊ Proactiveness and problem-solving: Proactive and self-disciplined members with problem-solving abilities contribute to efficient decision-making.

◊ Adaptability, teamwork, and diversity of perspectives: The ability to adapt and work harmoniously with diverse viewpoints and challenges fosters innovation and a comprehensive approach to problem-solving.

This analysis highlights the multifaceted nature of team composition in construction projects, emphasizing the importance of a holistic approach that considers technical skills, experience, interpersonal dynamics, and alignment with project objectives. When effectively combined and implemented, these factors form the foundation of a well-composed and high-performing construction project team.

The importance of adopting a systematic approach to team member selection cannot be overstated. Team effectiveness relies on the project team members, highlighting the critical importance of meticulous project team selection for overall project success. Team composition is one of the most prominent elements affecting trust among team members. Teams can ensure a balanced and effective composition by tailoring selection processes to project requirements, aligning individual skills with project demands, defining clear roles, and perpetuating objective evaluations. Emphasizing development, feedback, and diversity recognition enriches the team's problem-solving capacity and creativity. For instance, the response from I1, who had 25 years in cost controls and contract engineering in construction projects, echoed the literature almost exactly regarding the various steps that can be employed and deployed in the process of selecting and identifying an appropriate team. According to the interviewee, it is essential that the selection of team players in a construction project be first aligned and anchored on the requirements and the objectives of the project.

The identification of team members should satisfy the requirement of the project, and this should follow the line of merit of the individual's capacity to handle the needs of the project and deliver sufficiently. I3 also added to the discourse by mentioning that part of the success of the team selection process will be based on the ability and willingness to continually evaluate the effectiveness of the composed team during the stages of the project to find gaps in the team strengths and attend to them before they become an impediment to the success of the project. There was general agreement among the experts that the process of identifying a team has to be structured and strategic because it is only through this approach that the team can benefit from the expertise of each of the members and lead to the success of the project.

The foundational importance of qualifications and expertise emphasizes the necessity of educational qualifications, relevant project experience, administrative competencies, and alignment with project-specific needs. These elements ensure a robust knowledge base and streamlined activities to effectively meet project demands.

Zainal et al. (2020) found the role of a project team in the current highly competitive environment is to ensure that there is a higher productivity yield, there is growth in the innovative capability of the team, and that they are able to solve problems as they occur during the process of construction. To this extent therefore, Zainal et al. indicated that the level of consideration given to qualification, experience, and expertise during the process of identification and selection of a project team is commensurate to the level of success that the team is projected to achieve and accordingly to the level that it will attain. Likewise, I1 one mentioned that the candidate fitting a team should have a qualified and then related experience. I5 agreed. From a practical perspective, for a project team to be successful, then all the players in the team must have their own contribution that will build into the strength of the team and provide the team with the ability to succeed in its projects and endeavors.

The centrality of critical interpersonal skills for fostering collaborative team dynamics is also vital Effective communication, proactiveness, adaptability, and a spirit of teamwork significantly contribute to conflict resolution, innovation, and a comprehensive problem-solving approach. As Neto et al. (2023) argued, "It is possible to see that the success of projects is deeply linked to the balance between hard skills and soft skills, integrating appropriate leadership behaviors" (p. 15).

All this clarifies the need to review interpersonal skills individually. First, it is critical to conduct personality tests to examine the suitability of some people to work together in a perfect mix. This is reechoed from the primary data collected by I7 and I8. According to I7, the ability of a team member to understand other members of the team is paramount, and once every team member has this ability, it contributes to reduced friction and increased collaboration. I8 adduced that, in the team composition, the interpersonal skills are of great significance for instance, attitudes like commitment to tasks, collaborative readiness and skills, clear and well-developed communicational skills and overall compatibility with other team members are traits at the interpersonal level that are critical to the success of the team.

Clearly, project team members and leaders responsible for the selection and identification of project teams, pay attention to the application of the structured criteria that will help in the setting up of strong and robust teams. Such approaches can be put down in policy so that they can be referable and can guide the future identification, selection, and setting up of teams to ensure that the efficacy of the team is in-built right from the time of setting together the team. Project team leads should focus on developing policies around the selection criteria and strengthen these policies based on the premise of the importance of interpersonal skills and capabilities above qualifications and experience. This will lead to the establishment of effective construction project teams as far as the team composition element is concerned. Proper identification of team members greatly facilitates assembling teams that can work together on tasks and ensure the delivery of the task without usual team complications and conflict.

Chapter 6

Effective Team Building Overview

Having a supportive organizational culture and a well-balanced team composition are essential components, but that team must also be built up until it is fully developed to meet its goals. Team building can be achieved through the practical facilitation of a team approach or by engaging in formal and informal teambuilding activities. Team building focuses on enhancing work group skills, fostering social relationships, and resolving issues that can impede team performance. The essence of team building lies in creating the essential conditions for cultivating effective teamwork skills, identifying leaders, and fostering an atmosphere of informal communication (Sakalyuk, 2021).

Furthermore, the literature offered various team development interventions designed to enhance team proficiency, including team training, coaching, and debriefing. Shuffler et al.

(2018) argued that at a fundamental level, team development interventions must be applied to have any impact on a team. The authors reviewed different types of team development interventions in the academic literature and identified ten types that effectively shape team performance trajectories. These include team task analysis, team composition, teamwork design, team charters, team performance monitoring and assessment, team debriefing, team building, team training, team coaching, and team leadership.

In the pursuit of managing a thriving and efficient team, project leaders and firms are faced with a set of actions and techniques that they should embrace, as well as those they should steer clear of. Furthermore, project managers bear the responsibility of safeguarding their teams against both internal and external influences, continually assessing team performance, and embodying essential behaviors and attitudes to sustain team effectiveness. Team building is a complex and multi-faceted task that covers such issues as managing the team life cycle, training, and the need for continuous improvement over time. But to begin, I want to establish the elements of effective team building gleaned from the interviews and survey. There are three major factors, or goals, each with multiple team development techniques, as follows:

◊   Empowerment, Delegation, and Support

   ◊   Give space to the team members to do their job.

   ◊   Delegate with care and trust team members.

◊   Provide constructive feedback and recognition.

◊   Set a clear plan for the project.

◊   Empower your team to make smart decisions.

◊ Building Effective Teamwork and a Collaborative Environment

◊   Encourage unity, friendliness, and sympathy among team members.

◊   Foster trust, belonging, and inclusivity.

◊   Establish clear team boundaries.

◊   Maintain stability in team membership.

◊   Establish a framework for team operations.

◊   Create processes that support team success.

◊   Foster social connections by building camaraderie and trust among team members.

◊   Avoid negative comments and encourage positive interactions.

◊   Promote well-being, and foster physical and mental health at work.

◊   Establish strategies for identifying and resolving problems within the team.

◊ Comply with relevant standards and references for work execution.

◊ Providing awards or promotions for exceptional performance.

◊ Address wrong behavior within the team and applying warnings and penalties.

◊ Regularly checking in with team members to ensure their needs are met and they feel supported.

◊ Actively involve all team members and enhance team-build activities.

◊ Ensure accountability and establish clear rules and guidelines for responsibility.

◊ Address logistical needs for effective work operations.

◊ Conduct regular meetings to discuss work progress and issues affecting team performance.

◊ Prioritize tasks and providing guidance on task management.

◊ Lead through collaboration and guidance.

◊ Implement high organizational ethical standards to protect team members.

◊ Ensure a safe, healthy, and secure work environment.

◊ Provide the necessary resources, equipment, instruments, and tools for work.

◊ Motivate the team by mitigating weaknesses and supporting the strengths of each member.

◊ Personal Development

◊ Arrange technical workshops.

◊ Provide training and development tools.

◊ Avoid constant routine work to encourage individual skills.

◊ Mentorship for skill development.

◊ Embrace diversity and inclusion.

◊ Use 360-degree feedback.

◊ Build resilience and foster flexibility.

◊ Use active listening and empathy.

◊ Support positive attitudes.

◊ Encourage problem-solving behavior.

◊ Evaluate performance using standard criteria for all members.

◊ Empower positive behavior among team members.

◊ Integration of team members' expertise and skills.

◊ Encourage knowledge transfer among team members.

◊ Establish protocols for handling external influences.

◊ Establish accountability.

It's a daunting list, but these goals and techniques encompass a comprehensive array of strategies and actions essential for fostering team development as well as effective management of teams within the construction industry. These encompass various facets from giving space to team members to delegating with care and trust, providing constructive feedback and recognition, setting clear project plans, encouraging unity and sympathy, establishing clear team boundaries, establishing a framework for team operations, creating supporting processes, and fostering social connections and camaraderie. These components collectively contribute to the development of a robust, collaborative, and successful team.

Furthermore, the process of effective team building extends to promote positive interactions, promote well-being, and address problems proactively. These elements, when effectively executed, lead to the formation of a unified and high-performing team. In addition to these considerations, recognizing exceptional performance, addressing wrong behavior, regular check-ins, ensuring accountability, conducting regular meetings, prioritizing task management, implementing ethical standards, ensuring a safe work environment, and leading through collaboration are integral aspects that contribute to the construction of a robust and ethically sound project team. Other elements include providing necessary resources, motivating by leveraging strengths, arranging technical workshops, providing training and development, embracing diversity and inclusion, using 360-degree feedback, and establishing protocols for external influences.

I will cover more from the interviewees in the next few chapters, but for this overview, I'll start with some comments about team member empowerment, delegation, and support. Again, there was an emphasis on delegating decision-making, with I1 saying leaders should give space to the team members to do their job, empower the team to make smart decisions, and provide constructive feedback and recognition. For building effective teamwork and a collaborative environment, many discussed the need to take care of the team as a whole and individually. I7 said to be sure that team members receive their salaries on time, I5 talked about making sure the team has the resources it needs for its tasks, and several talked about being friendly and creating a friendly, supportive environment. As I3 said, "Establish and maintain the corporate culture and positive behavior among the project team." I8 advised checking in with teams regularly, and I 2 stressed the need to support team members in their progress. Almost all addressed the need to establish clear goals and interpersonal trust. Finally, for now, regarding personal growth and character development, I1 I5, and I8 talked about training and workshops, I8 talked about encouraging problem-solving behavior and keeping members accountable for their work.

During the interviews, it became evident that participants widely recognized the organization's profound emphasis on key areas such as communication, diversity, and inclusion. These cultural factors significantly impacted team dynamics and contributed to a cohesive sense among team members. The commitment to fostering an inclusive environment was perceived as a cornerstone of effective team management. Additionally, the importance of selecting team members with the right skills emerged as a unanimous consensus among interviewees. This strategic approach to team composition was viewed as pivotal for enabling effective problem solving and decision-making within construction teams. It was clear that the organization's focus on assembling teams with the appropriate skill sets positively influenced overall team management and performance. A recurring theme highlighted by many participants was the value placed on regular problem-solving meetings. These meetings were deemed indispensable for enhancing team collaboration by providing a platform to address challenges and brainstorm solutions, thereby maintaining a productive team environment.

Moreover, the interviews emphasized the significance of clear career development strategies within the organization. Workshops, seminars, and various learning opportunities were considered invaluable assets for team members seeking to enhance their skills and advance their careers. The organization's commitment to facilitating continuous learning and growth was widely appreciated. Similarly, the participants positively perceived the availability of professional and management training programs instrumental in equipping team members with the necessary skills and knowledge to excel in their roles. This proactive approach to providing skill development and career advancement opportunities was seen as a testament to the organization's commitment to its employees. Interviewees also valued the emphasis on delegation as a technique for enhancing team skills and promoting autonomy within the team. Empowering teams to make decisions and exercise autonomy was seen as fundamental for promoting ownership, accountability, and motivation among team members, thereby driving innovation and problem-solving. Clear performance evaluation criteria were deemed important by interviewees as they enable team members to understand how their contributions are assessed and aligned with organizational objectives, promoting fairness and transparency essential for team morale and motivation.

Chapter 7

Managing The Team Life Cycle

The experts I interviewed identified managing the team life cycle in the construction project teams as one of the elements of effective project management, so I studied it in-depth and found that effectively managing the life cycle consists of six categories: effective communication, team environment, team structure, goal setting, establishing clear roles, and setting expectations.

I should note here that during the validation of the maturity model developed from the qualitative phase of the study, I relied on eight items of measurement. However, one of these items was unfit, and I redacted it from the final maturity model. In the review of the literature, the research had found out that the management of the team life cycle was made up of sub-elements such as understanding the different stages of team life cycle development, problem-solving skills, conflict resolution, and management skills, among others. According to evidence from extant literature, this is one of the constructs in the model that was specifically for the team leaders and not for all members of the team (Vaida & Serban, 2021). The highest-ranking measurement item for the team life cycle management was that planning and task distribution is essential for construction management. This item of measurement scored a very strong loading against other elements in the model, and this confirmed the sentiments gathered from the qualitative phase of the study as well as those gathered from the literature review.

It is essential that project management team leads prepare proper structured instructions and tasks to guide the deployment of efforts by the team members to ensure that there is no wastage and that all efforts are directed toward the goal of the project team. This promotes success because there is a clear understanding of the roles and responsibilities of all the team members and how they are supposed to contribute to the achievement of the overall goal of the organization. Other items of measurement that were used to measure the construct of the management or project team life cycle included goal setting, clear roles and expectation; team performance management; team environment, and structure. The item of measurement of goal setting scored a high loading in confirmation of the importance of clarity of the project during the execution. Conflict management and dispute resolution are essential for construction team management, which is a testament to the fact that in the management of the construction project life cycle, the aspect of dispute resolution does not carry so much weight. Accordingly, therefore, it is recommended that in the management of the project life cycle, proper attention be paid to the aspects of clarity of instructions and goal setting to steer the team forward in an objective and clear manner.

A project team is essentially a social construct, where members influence each other and are shaped by their relationships throughout the project's progression. Thus, "as teams' form with the aim of achieving a particular goal or higher-order purpose, it is only logical to anticipate that they will evolve, grow, and develop over time as they collaborate towards their objectives" (Shuffler et al., 2018, p. 21). Consequently, "a genuine team doesn't materialize overnight but must traverse several stages to transcend its initial existence as a mere group" (Vaida & Serban, 2021, p. 241). Therefore, understanding the stages of team development or its life cycle is paramount for the team's success and effective management. "Attending to group development can significantly enhance the group's performance and, ultimately, contribute to what is commonly regarded as project success" (Gren et al., 2016, p. 3). Recognizing the stages of team development thus empowers the team leader to employ an appropriate leadership style tailored to each stage of development, so it's hardly surprising Natvig and Stark (2016) argued that "a leader proficient in guiding the team through the phases of group development is indispensable" (p. 680).

Construction practitioners have highlighted performance evaluation as one of the crucial elements for the management of the team life cycle that contributes to the effective management of the team.

"The performance evaluation system is an important measurement tool used as a support mechanism or various areas of the organization, aiming to analyze the performance of professionals, ensuring the balance of the current reality versus the organization's planning" (Guilhermeet al., 2019). Islami et al. (2018) said that "wanting people to perform at a high level, high standards of performance should be set" (p. 94). The authors' findings are remarkable in that "individual performance evaluation explained 41.2% of employees' productivity" (p. 105).

We know that performance evaluations greatly impact workers' motivation and development. But further, "When measuring the performance of team-based structures, it will be more productive to carry out measurements and evaluations in a hierarchical and multi-dimensional manner, thereby enabling performance to be managed efficiently" (Silva et al., 2021, p. 100). Moura et al. (2021) contributed to the discussion on the factors influencing the high performance of project team members and showed "a set of perceived factors that can influence (facilitate/hinder) the high performance of IS project team members. "Proper reward systems" was the most mentioned facilitating factor; "negative affectivity" and "lack of competence" were the two most prevalent hindering factors."

In reality, the demarcation between these developmental stages is often blurred. It's possible that just when it seems the team is progressing along its developmental path, it may "regress" or "revisit" earlier stages in response to new demands, such as the arrival or departure of a team member, changes in work processes, or other factors influencing team dynamics (Vaida & Serban, 2021). Effectiveness develops over time as groups come to understand what is expected of them and how to fulfill those expectations.

In developing a life cycle for this book, I was fortunate enough to draw from other models of team development in the literature, include Bass and Ryterband's 1979 model (acceptance, communication and decision-making, group solidarity, group control), Tuckman's 1965 model (forming, storming, norming, performing, adjourning), and Cog's Ladder (Polite, Why we're here, Power, Cooperation, Esprit de corps).

These models have been used extensively in the literature. Jones (2019), for example, applied Tuckman's model to a small group of students, and the research findings highlighted that the model serves as a valuable tool for tracking progress, evaluating skills, behaviors, emotions, flexibility, adaptability, and adjustment to varying circumstances, cultural interactions, and diverse case scenarios. At a micro level, it enables the observation of behaviors, problem-solving capabilities, critical thinking skills, management proficiencies, and leadership qualities.

Authors have argued that there is no one superior or universally better model. The choice depends on the specific requirements of your team, your leadership experience and expertise, and the available resources

Boiled down, from all these models we can see the influential factors of the team life cycle:

◊ Understanding the stages of team development

◊ Effective leadership style suits each stage

◊ Observe behavior

◊   Problem-solving skills

◊   Conflict resolution

◊   Management skills

Looking at the core element of all these factors, the common element of the team life cycle that contributes to the effective management of the construction team is clearly good communication. I'm hardly the first to notice this, and many books and articles before this one have heralded the role of effective communication with and among team members as an anchor of success in the process of team engagement.

Gamil and Rahman (2017) summed it up well:

> the construction industry is characterized in nature as complex, fragmented, dynamic and involves many parties therefore effective communication is essential to overcome these challenges . . . many researchers found that the industry faces a major challenge to ensure effective and successful communication throughout the life cycle of the project which therefore resulted to project failure. (p. 239)

Indeed, Islami et al. (2018) found that "each 1% increase in communication will predict an increase in employees' productivity by 24.8%" (p. 105). Therefore, effective communication among team members' fosters collaboration and problem-solving and leads to project success.

The experts I interviewed agreed and shared in depth some of the challenges that construction teams phase due to poor communication. They told me one challenge that lack of effective communication may lead to the poor delivery of the project owing to the fact that the deliverables are not clear among team members and this leads to inconsistencies in the process of implementation of ideas, which then lead to poor project outcomes. To counter this, and in an effort to underline the importance of effective communication as far as the management of effective construction project teams is concerned, the interviewees noted that project team leaders must master the art of allowing open communication among members by leading the communication with openness.

I1, I2, I4, I8, and I9 all called for being "open" with the team, with I9 stressing the importance of "clear and honest communication." Regular communication was cited as vital by I1, I2, and I5. Project teams that communicated regularly were able to attend to problems that may occur in the organization in advance before the occurrence of the problems. most talked about the goals of communication as building trust, clarity, and empathy on both sides. I3 stressed this meant talking with different members separately.

There was general agreement among all participants that the project team's life cycle depends on effective communication as one of the core pillars without which, the project team cannot effectively coalesce around an idea and deliver it because there is a total breakdown. From a practical perspective, both findings from extant literature and those from primary research and the convergence between the findings of the absolute importance of effective communication in the successful management of project team's life cycle, are valid and can be justified in any project instance. It is known practically as noted from literature and confirmed from the primary research that project teams that have effective regular communications would easily deliver a complete project and that one of the impediment of projects completion or one of the antecedents of poor projects is poor communication. To this end therefore, effective communication should be handled as of paramount importance in the successful management of the project team's life cycle.

Concerning the vital skills of conflict management and dispute resolution, all said that conflict must be recognized as inevitable and dealt with directly, and most espoused that the best way to do this was to find the "core causes" of such conflict. I6 called for "Identifying points of agreement and disagreement," and I9 noted the importance of resolving conflicts early on. I8 talked about "Collaborating with team members to find solutions and prevent reoccurrence." I5 suggested selecting a popular team leader to resolve internal disputes and then involving management then the conflict couldn't be resolved that way. I2 said the team's shared vision should be used to find common ground, and I6 advocated focusing on behavior and events rather than personalities. All took on conflict as a time for problem-solving, and I6 said they would build on the success of each conflict resolution.

Regarding issues of team environment and structure, almost all talked about fostering a friendly, open, and collaborative atmosphere where members felt they could trust each other and their leaders. I5 stressed basic shows of respect like being punctual, and I8 encouraged getting to know team members personally. I8 also encouraged being patient and flexible.

In discussing setting clear goals, roles, and expectations, almost all stressed the need to be clear from the very beginning and making sure all team members knew their own roles and responsibilities for what I1 called a "clear task structure." I8 stressed setting clear accountability, and I9 pointed to clearly defining a project's scope, job descriptions, and timelines. I6 discussed the need to check in with the team regularly to ensure progress was being made, and I4 said to use KPIs to measure that progress. As I1 put it, it is essential to "Quantify what success means," and I1 advised to revisit project goals regularly.

For planning and task distribution, I1, I2, and I6, stressed the importance of making sure everyone knew their role and the general distribution of tasks. I1 mentioned the need to use good management software, and I8 said to check in regularly with the project manager. Regarding performance management, almost everyone advocated for continuous monitoring and evaluation, or what I6 called "regularly tracking and recording." Providing feedback, listening carefully to the team members, and conducting performance reviews were also emphasized. I1 advised taking note of "negative and positive outliers."

The journey of a team, from its inception during the forming stage to reaching peak performance, often involves confronting numerous challenges and obstacles. As a result, the adept management of the entire team life cycle becomes a pivotal factor in ensuring optimal team performance. we present an intricate breakdown of themes and sub-themes thoughtfully identified by seasoned construction practitioners. These insights shed light on the critical elements for effective management of construction teams, particularly in terms of managing the team's progress from the post-selection phase through its developmental phases until its eventual disbandment, all aimed at achieving the pinnacle of effective performance.

Chapter 8

Effective Team Management And Training

According to Ginting et al. (2020), "Training is a structured process designed to instigate changes in attitudes, knowledge, or skills through experiential learning, with the aim of achieving effective performance across a range of activities" (p. 5). Also, "It is widely believed that when teams are employed effectively and equipped with the necessary training, organizations can attain their anticipated outcomes" (Al-Malki & Juan, 2018, p. 53), and "Among all the research on team development interventions, the evidence supporting team training is arguably the most robust" (Shuffler et al., 2018, p. 12).

Regarding training methods, four strategies are generally employed to enhance the regularity and efficiency of team performance and teamwork processes. These four methods comprise "the delivery of didactic lectures/presentations, workshops, simulation exercises, and on-site review activities" (McEwan et al., 2017, p. 3). The authors further conducted a comprehensive systematic review and meta-analysis of teamwork interventions aimed at enhancing teamwork and team performance using controlled experimental designs. Their findings indicated noteworthy and positive effects of team training interventions on team processes and performance.

Teamwork training demonstrated its effectiveness in enhancing teamwork and team performance across a diverse range of team contexts. This observation was substantiated by (Salas et al., 2008), who stated that

> team training interventions were more effective for team processes, and it is logical to assume that enhancing processes will positively influence performance outcomes … Implementing effective team training best practices led to improvements in learning, team cognition, emergent team processes, and performance. (p. 926)

Umar et al. (2020) revealed that "practical training could enhance employees' soft-skill competence, foster employee creativity, and boost team performance. Moreover, the authors argued that training plays a significant role in enhancing employee creativity and optimizing team performance" (p. 1021).

Lacerenza et al. (2018) asserted, "Team development interventions play a critical role in enhancing effective teamwork competencies and processes, ultimately leading to improvements in both proximal and distal outcomes" (p. 517). Their research delved into team training, leadership training, team building, and team debriefing to examine the impact of these interventions on the team's dynamics. Team training, simulation, coaching, and debriefing are vital components of team expertise." Smolska (2022, p. 512) explored the characteristics of project teamwork, the legitimacy, and the frequency of employing developmental methods, which encompass training, consulting, coaching, and mentoring, within project teams. The author unveiled, "The choice of development methods for a project team should be contingent on the specific project team, its unique characteristics, and the nature of the work. Different project teams with diverse needs may necessitate tailored sets of development methods." The author underscored the importance of continuously diagnosing the development requirements of both individual team members and the team.

The effects of teamwork training on teamwork outcomes are significantly more pronounced for newly formed teams in comparison to existing teams. This discovery contrasted with Salas et al. (2008), who argued that team training worked equally well, if not better, for intact teams, especially concerning performance outcomes.

While all four training methods (didactic lectures, workshops, simulation, and review-type activities) proved effective in enhancing team performance and outcomes, interventions centered on didactic instruction did not yield significant improvements in teamwork itself. As Salas et al. (2008) pointed out "For process outcomes, task-focused training was less effective than teamwork training. Similarly, training that emphasized teamwork appeared to produce enhanced affective outcomes compared to task-focused training" (p. 923).

The interviewees had a great deal to say on training, though they often used other teams such as "coaching," "fostering," "enhancing skillsets," and "gaining confidence." They stressed that while individual skill-building is important, the training overall should lead the group to a "shared mindset," (I1) and one should "Encourage activities that strengthen relationships between team members" (I4). I8 stressed that the point of "providing adequate training for knowledge and skill development" was so employees should "look for opportunities to improve skills and knowledge."

The interviews also clarified that training should be part of encouraging continuous learning at the individual and team levels. Luo et al. (2023) found that "Vision guiding and context interacting have both direct and indirect effects on the project management performance" (p.1). The authors also discovered that "team building can improve the project management performance by influencing project citizenship behavior" (p. 16) and stimulate learning and personal development. Thus, leaders who prioritize learning contribute to the growth and adaptability of the team in the ever-evolving landscape of construction projects. Thus, recognizing and addressing the personal needs of team members, promoting work-life balance, and encouraging a positive environment contribute to the overall well-being of the team.

A leader must deal with the psychology of those on their staff and take on their perspective. Jena et al. (2018) stated that "Executives have a higher level of trust when they can able to perceive meaningful employee engagement, transformational leadership and psychological well-being from their respective employer" (p. 227). Therefore, leaders prioritizing these factors will likely retain a motivated and engaged workforce.

Chapter 9

Team Leadership

A team leader should exhibit behaviors that foster the development of a unified team and a productive work environment, ultimately leading to the desired outcomes. To exert a positive influence on the team's performance, team leaders must undertake several crucial steps, including setting clear team goals, delineating specific roles for team members, effectively communicating the team's purpose, resolving conflicts, and nurturing cooperation within the team. As Al-Malki and Juan (2018, p. 40) asserted, "The management of teams and the assurance of their effectiveness hinge on the presence of leaders who can inspire and provide a vision for their subordinates." Grynchenko et al. (2018) argued that "the effective formation of the team requires a leader that must have the relevant knowledge and skills, strategic thinking, psychological culture and the ability to assess the professional competence of candidates and their significant professional and social personal qualities" (p. 13).

Furthermore, team leaders should articulate a compelling vision that motivates employees to assume greater responsibility for their work. Providing clarity on goals and specifying tasks, roles, and rewards, understanding employee needs, cultivating a supportive atmosphere, and engaging in confidence-building practices that can significantly enhance performance. Additionally, leaders and their management styles play a pivotal role in managing role conflicts and identifying the primary sources of role-related stressors. For example, mentoring, a function found in various leadership styles, plays a crucial role in addressing role stressors (Al-Malki & Juan, 2018, p. 57).

Setting clear goals and defining roles and expectations are essential for motivating employees and ensuring project success in construction teams. "Communicating objectives and setting clearly the goals that employees must realize, results in reaching them by working with satisfaction by employees" (Islami et al., 2018, p. 94). The authors further concluded that "when the employee knows precisely what he or she is going to be reached with the project that he or she is part of, the employee will work with more motivation and high concentration to realize that project" (p. 105–106). Epton et al. (2017) explored the distinct impacts of setting objectives on altering behavior, spanning various fields, demographics, and situations. The authors revealed, "Setting specific and difficult goals effectively increases behavior change. However, there was no evidence to support that task complexity, feedback, and commitment boost the effects of goal setting. Instead, their findings show that goal setting is optimally effective when "(a) it is set face-to-face, (b) it is set publicly, (c) it is a group goal, and (d) it is coupled with monitoring of the behavior or outcome by another person without feedback" (p. 35).

Clear direction in the team is one of the pillars of the longevity of a project team life cycle. The experts I interviewed noted that goals, roles, and expectations enabled the team members to understand their contribution to the team and own this contribution. From a practical standpoint, once the team members, through effective communication and operating in a positive team environment, understand their goals, roles and expectations, it would be very fluid for them to own up to their roles and expectations and contribute to the teams' successes from a point of commitment, ownership and team membership.

Effective planning and task distribution are vital for meeting project deadlines and budget constraints in construction teams. Liu et al. (2019) said that "completing a construction project on time and within budget is important in the construction industry. To achieve this goal, a construction plan satisfying the time and cost constraints is crucial" (p. 472). Islami et al. (2018) suggested that "creation of controlling list serves as a tool that is used to maximize employees and managers' efficiency, and serves as a pusher to increase the level of finished duties" (p. 106). The authors further concluded, "Flexibility in roles can be beneficial, but too much variability can be associated with role ambiguity and negatively affect a team's ability to achieve its goals." Loberg et al. (2018) linked the forced distribution rating systems (FDRS), as performance appraisal, with the task. They found that "FDRS significantly increases team productivity if task interdependence is low. If task interdependence is high, however, an FDRS significantly decreases team productivity, even if sabotage is unlikely to occur."

The interviewed experts agreed that planning strategies like introducing clear action plans for the execution of tasks that can be followed by all team members, delegating tasks based on the strength of team members, using good project management software that enhances collaboration and information sharing and regularly reviewing the work plans and its validity are critical in ensuring there is success and longevity in the project team life cycle. These sentiments when considered practically and juxtaposed by the evidence from literature, it is clear that although autonomy in the execution of tasks is something to be pursued and has been identified to lead to greater commitment on the part of team members, the function of the team leader should be to set up a clear plan and ensure that this plan can be adhered. Sticking to a plan has been noted by both extant literature and primary research as one of the approaches through which projects' successes can be measured and in the practical aspect, this is very valid.

Team leader behavior serves as a critical catalyst for creating the necessary conditions to foster team learning behavior. In their research, Ivanov et al. (2021, p. 276) outlined nine characteristics of effective leadership, which encompass soliciting cooperation, avoiding unnecessary competition among team members, practicing transparent communication, demonstrating strong planning skills, generating innovative ideas, displaying flexibility, being accountable, and showing care for team members. Thus, it is evident that leaders play a direct role in promoting the conditions necessary for the success of their teams (Paolucci et al., 2018, p. 137).

It is evident that a single leadership style is not universally effective in all situations. Thus, effective leaders possess the ability to adapt their leadership styles to suit the atmosphere and the specific work environment, showing an innate understanding of what is required and the individuals they are leading. The choice of leadership style in a given scenario hinges on various factors, including the stability, conservatism, or dynamism of the environment. Therefore, leaders exhibit a range of multifaceted behaviors and adapt their leadership styles depending on the specific context, as well as in response to the demands associated with decision-making, the prevailing situation, and the characteristics of their followers.

Martin et al. (2018) identified shared leadership typology, personality composition, work complexity, and types of team performance as moderating factors of leadership. Understanding these factors can empower project managers and their teams to enhance overall team performance. The authors concluded that "shared new-genre leadership (NGL) exhibits a stronger correlation with team performance than shared traditional leadership. This suggests that teams with growth-oriented approaches are more likely to excel than transactional teams focused on maintaining the status quo" (p. 677).

Choi et al. (2017) delved into the study of transformational leadership as a top-down style stemming from the formal leader of a team and shared leadership as a decentralized style emanating from team members. Their findings revealed that transformational leadership significantly contributed to team output effectiveness, while shared leadership enhanced the team's organizational and planning effectiveness. These results imply that distinct leadership styles have varying impacts on various facets of team effectiveness. The authors recommended that managers collaborate more closely with team members and align the leader's behavior with the characteristics of the team's goals to promote overall team effectiveness. Paolucci et al. (2018, p. 140) conducted research on the "effects of transformational leadership on three dimensions of team effectiveness: viability, team process improvement, and the quality of the group experience." Their study revealed that leaders adopting a transformational leadership style were able to generate positive outcomes in team viability and the quality of the group experience, although the team process improvement dimension did not exhibit statistically significant improvement. Furthermore, their results suggested that affective team commitment played a partial mediating role in the relationship between transformational leadership and the quality of the group experience. Therefore, it is imperative that those involved are cautious in idealizing particular leadership forms. Furthermore, it is important to recognize that leadership and team motivation are linked. Effective leadership is closely associated with the sustained motivation of team members. Smith (2020) contended, "Prominent leadership styles such as transformational, transactional, authentic, and servant leadership influence team motivation. Specifically, transformational, authentic, and servant leadership styles exhibit a positive correlation with team motivation, while the transactional leadership style shows a negative correlation" (p. 726).

From the interviews and survey, I found nine major team leadership factors, each with multiple techniques, as follows:

◊ Emotional Intelligence and Team Dynamics

    ◊ Managing emotions effectively

◊ Building solid relationships with team members

◊ Demonstrating empathy and understanding as a leader

◊ Recognizing and addressing the personal needs of team members

◊ Modifying relationships based on empathy and understanding

◊ Team Inclusivity and Diversity

  ◊ Creating an inclusive and diverse team

  ◊ Acknowledging the importance of diverse backgrounds and perspectives in a team

◊ Recognition and Appreciation

  ◊ Valuing each team member's contributions and ideas

  ◊ Fostering a culture of appreciation

  ◊ Recognizing and appreciating the achievements of the team members

  ◊ Promoting team members based on task achievements

  ◊ Leadership and Guidance

  ◊ Being confident, inspiring, and guiding the team to success

  ◊ Leading by example in a collaborative setting

◊ Encouraging team members to exhibit leadership capabilities

◊ Goal Setting and Communication

◊ Setting clear goals and direction

◊ Efficiently communicating goals and expectations

◊ Sharing a clear vision and setting goals for the team

◊ Continuous Improvement and Problem-Solving

◊ Encouraging continuous learning and development

◊ Encouraging a culture of continuous improvement within the team

◊ Providing practical solutions to address daily work issues

◊ Continuous feedback and adjustment for improvement

◊ Team Motivation and Well-Being

◊ Delegating tasks to team members to maintain motivation

◊ Meeting the needs of team members

◊ Encouraging work-life balance for team members

◊ Striving for excellence and innovation

◊   Collaboration and Trust

   ◊   Emphasizing the need for teamwork and collaboration

   ◊   Building trust with the team through open communication and a positive environment

   ◊   Empowerment and Ownership

   ◊   Empowering team members to be the initiative

   ◊   Empowering team members to take ownership

   ◊   Encouraging self-assessment and self-correction

   ◊   Promoting a sense of accountability within the team

Creating an environment conducive to effectively managing the project team requires careful consideration of several crucial factors. Addressing issues and conflicts that may arise among team members, striking a balance between the collective needs of the team and the individual needs of its members, and sustaining motivation among team members can pose formidable challenges for any leader. Consequently, the process of creating, nurturing, and preserving an effective team entails a continuous commitment and a readiness to invest in the team's triumph.

The interviews yielded several insights concerning team leadership qualities and responsibilities. Most addressed issues of emotional intelligence in leadership. I3 and I4 specifically discussed the need to show empathy, I1 talked about supporting mental and emotional well-being and "fostering a culture of appreciation." I2 discussed "empowering team members to take ownership," which I7 called "encouraging team members to exhibit leadership capabilities.

Of course, communication was mentioned often, particularly in regard to goals and expectations. Many talked about the need to be supportive and to build trust. I9 focused on diversity, inclusivity, and equal participation by team members. I4 mentioned delegating as appropriate, and I1 encouraged work-life balance. I8 said leaders must be "regularly checking in with team members to ensure their needs are met and they feel supported." Setting clear goals and providing efficient communication are foundational aspects of effective construction leadership. Leaders who articulate a clear vision and expectations, coupled with continuous feedback and adjustment for improvement, create a focused and responsive team.

Ssenyange and Kudakwashe (2023) concluded, "When project leaders allow stakeholders' participation and communicate effectively about project tasks, goals, strategies, and processes they are able to enhance the stakeholders' levels of commitment, trust, and cooperation enabling them to execute timely, cost-effective, and quality government construction projects." (p 11). Luo et al. (2023) stressed that, "under the pressure of the epidemic, leaders should increase communication with participants to build trust to enhance organizational leadership, thus promoting the project management performance" (p. 1). Kuzior and Balahurovska (2023) considered the "formation of effective communication with subordinates is one of the essential tasks of an effective leader and such communication can provide a solution to an essential managerial task—understanding the socio-psychological characteristics of the team" (p. 179).

While interviewees sometimes used "training" and "coaching" interchangeably, the latter concerns the leader, whereas the former may not. Effecting behavioral change, enhancing performance, or cultivating effective teamwork is not a singular achievement that can be exclusively attained through training interventions. As elucidated by Passmore and Gordon (2018), "While training may temporarily equip individuals with these skills, their ingrained habits are likely to resurface." Consequently, coaching emerges as a vital process that contributes to fostering behavioral transformation. Moreover, coaching is recognized as one of the most potent tools for enhancing individual development and team effectiveness. "Coaching has found widespread application within organizations, serving diverse purposes and yielding a variety of outcomes" (Passmore et al., 2019, p. 4). It stands as a pivotal organizational strategy for effecting behavioral change, nurturing skill development, motivating individuals, applying problem-solving techniques, and employing interventions to enhance team performance. As highlighted by Smolska (2021)

> In contemporary times, coaching plays a pivotal role in both individual and team developments. It empowers individuals and teams to unlock unexplored potential, set ambitious goals, fuel motivation, and overcome external and internal obstacles. Coaching enables individuals and teams to evolve into their best selves, encompassing personal growth and self-realization. Furthermore, coaching extends beyond individual objectives, making it a valuable method for fostering group and teamwork within enterprises. (p. 1)

Individual coaching is a personalized process that involves one-on-one guidance and instruction aimed at enhancing an individual's knowledge, skills, and job performance. Woo (2017) stated coaching is typically employed to help individuals uncover their full potential and optimize their performance. Many a book and management seminar has underscored the positive impact of coaching, including improvements in job performance, the development of self-awareness, increased employee satisfaction, contentment with work, and satisfaction with one's manager.

Romão et al. (2022) examined employees' perception of their leaders' coaching skills and found that "leaders' coaching skills have a negative impact on employees' turnover intention and a positive impact on their happiness, with the latter mediating the relationship between the first two variables" (p. 1). The authors recommended that organizations should recognize the utility of a leader's coaching skills, not only for enhancing employee happiness but also for mitigating turnover intention. Consequently, organizations should encourage leaders to incorporate more coaching skills into their leadership styles. Coaching serves as a tool for cultivating customer-oriented employees and supporting employees in identifying opportunities for improving their job-related skills. Coaching supervisors, according to the author, foster a team approach that emphasizes power-sharing, serving others, and teamwork. This approach can heighten affective commitment to supervisors due to increased cohesion among team members and a sense of being valued.

There are clearly positive relationships between managerial coaching and role clarity, job performance, and organizational commitment. HR development strategy should emphasize managing high-quality human capital through coaching methods to enhance organizational performance. Therefore, "a competent manager with coaching abilities can mitigate the effect of perceptions of organizational politics, further fostering thriving at work and job performance" (Raza et al., 2018, p. 30).

Ali et al. (2020) concluded the following:

> The role of a manager as a coach is pivotal in enhancing the feedback orientation of subordinates, a crucial factor for effective employee performance. This cultivates a relationship between supervisors and subordinates characterized by open communication, allowing dyads to freely exchange ideas, emotions, and concerns. Managers actively foster a team-oriented ethos among employees, a valuable asset in modern organizations. They prioritize the needs of subordinates in the workplace and facilitate employee learning beyond traditional roles of direction and control. These managerial behaviors result in heightened affective commitment to the supervisor, benefiting all stakeholders, including employees, managers, and the organization. Employee affective commitment stands as a highly sought-after form of commitment for both supervisors and organizations, as it leads to improved employee outcomes within the organization. (p. 17)

Furthermore, Sharma (2017) pointed out that

> coaching may yield various outcomes at different phases of the coaching journey. Initially, intangible personal

outcomes are realized, which subsequently facilitate the emergence of more tangible outcomes. This mediating role of intangible personal outcomes underscores the equal importance of both intangible and tangible coaching results. Consequently, assessing coaching effectiveness should place equal emphasis on collecting data on intangible and unintended outcomes, in addition to tangible results. This suggests that multiple criteria for evaluating coaching effectiveness may be necessary to comprehensively capture the full contribution of coaching in organizations (p. 14).

In a similar vein, Ali et al. (2020) argued that coaching should stimulate innovative behaviors and promote workplace learning through daily interactions between subordinates and managers. Their research delved into the direct and indirect influences of managerial coaching (MC) on innovative work behaviors (IWBs), which revealed a positive relationship between MC and IWBs. The findings underscored that MC has evolved into an effective managerial practice, an organizational development strategy, and a cost-effective training method for enhancing employees' skills to thrive in the dynamic and complex corporate landscape. When managers embrace their role as coaches, they prioritize the individual needs of their subordinates over organizational tasks, creating an environment where subordinates can easily exchange ideas with their supervisors. Additionally, this approach fosters a team-oriented culture within the organization, offering individuals more opportunities for learning and development as active team members.

In spite of the growing body of literature pertaining to team development interventions, the concept of team coaching is relatively recent. Jones et al. (2019) defined *team coaching* as a team-based learning and development intervention that treats the team as a system and is collectively applied to enhance team performance and achieve common or shared goals (p. 75). Team coaching involves specific activities that foster self and team reflection, facilitated by team coaches who employ coaching techniques like impactful, reflective questioning to raise awareness, build trust, and improve communication. This process also entails observing and interpreting dynamic interactions and is typically delivered over a series of sessions, rather than as a one-time intervention. De Vries (2020) emphasized that team coaching elevates team dynamics by addressing elements such as shared team or organizational purpose, enhanced collaboration, breaking down silos, promoting open communication, fostering creativity, and increasing the team's external focus.

At its core, team coaching is a learning process aimed at enhancing performance in relation to a team's shared goals. This journey of learning implies change, which may involve individual shifts in thinking, consensus on collaborative actions, and influencing the broader business, political, or social context that pertains to the team's organization, customers, and stakeholders. Team coaching has a positive impact on team performance processes, including effort, skill, and knowledge, which in turn directly contributes to team effectiveness. When team coaching is implemented without organizational constraints and at the right stage of a team's development, it can lead to significant enhancements in the team's overall effectiveness. The authors further noted that team coaching distinguishes itself from team building by focusing on performance outcomes, differs from team training due to its systemic approach, and is less structured than team facilitation.

Widdowson et al. (2020) referenced the best practices in team coaching, as outlined by Peters and Carr (2013). These best practices encompass the following:

◊ Assisting leaders in shifting their focus toward the initial team design and launch, rather than attempting to redirect a team once it is already in progress.

◊ Timing interventions to coincide with the start, middle, and end of a team cycle.

◊ Conducting a team launch, while being cautious not to overly structure the group or provide excessive guidance during the initial session, allowing the team to chart its own course.

◊ Encouraging teams to invite members to assume informal peer coaching roles within the team, motivating and inspiring their colleagues to contribute their full potential.

◊ Recognizing that team coaching, although primarily team-focused, can also encompass specific individual coaching for team leaders or members.

Furthermore, De Vries (2020) highlighted the advantages of team coaching, which encompass the following:

◊   Fostering a deeper awareness of shared circumstances, offering the chance to glean insights from others' experiences.

◊   Enhancing comprehension of how an organization shares, processes, and utilizes information.

◊   Potentially boosting productivity by promoting the adoption of shared practices and the application of lessons learned.

◊   Providing an avenue for coaching while allowing participants to draw from the successes and challenges of their fellow team members.

◊   Cultivating increased self-mastery and leadership skills as each participant offers support and coaching to their colleagues.

◊   Generating a synergy of energy, commitment, and enthusiasm due to the immersive nature of the coaching process.

Of utmost significance, team coaching is highly scalable, rendering transformative changes within organizations more attainable.

To address the question of whether effective team coaching could alleviate performance deficiencies in the workplace, Salihovic (2021) conducted a comprehensive review of the literature on team coaching. His conclusion emphasized that team coaching plays a pivotal role in establishing a new level of relationships within teams and organizations. This empowerment creates an invisible force that propels collective progress, leading to tangible differences and adding substantial value to a company. Notably, team coaching achieves this by boosting optimism, competence, and overall performance.

The benefits of team coaching extend beyond the immediate team, encompassing team leadership development as well. Fontannaz and Cox (2020) conducted an 11-month case study to delve into how team leadership development is nurtured over time through a dynamic coaching intervention. Their findings underscore several key points, as follows:

◊ Coaching supports team leader development during transitional phases, enabling leaders to engage with their experiences and bridge any gaps or tensions between expectations and reality.

◊ Coaching assists leaders as they assume their team leadership roles by providing both support and challenges.

◊ Continuous coaching contributes to the establishment of a coaching culture within teams.

◊ Coaching considers as a vital factor in fostering confidence, acting as a mediator in the relationship between leader development and team performance.

While the interviews did not address team debriefing in detail, several of the factors involved in debriefing were touched upon, so I want to explain its role in coaching.

Bisbey et al. reported, "While team training and coaching are often formal team development interventions led by trained facilitators, team debriefs are often led by team members" (p. 197). Although debriefing is used intensively in the health sector, aviation, education, and the army, "it plays an imperative role in the process of professional development in various fields" (Ku et al., 2019).

Lyons et al. (2015) explained,

> A team debriefing constitutes a structured and guided conversation among team members that occurs after a specific action period, serving as a platform to review and reflect on team performance. During these discussions, team members share their perspectives on what transpired, why it happened, and how they can enhance their collective performance. (p. 115)

The primary objective of debriefing is to "identify and rectify gaps in knowledge, skills, attitudes, and communication, which relate to individuals, the team as a whole, or the broader system" (Decker et al., 2021, p. 27).

Effective debriefing results in team members' attaining a shared understanding of their respective roles, team priorities, strengths, and weaknesses. Furthermore, team debriefs facilitate the development of "shared mental models, collective knowledge structures encompassing task, and team-related knowledge proven to enhance overall team effectiveness" (Lacerenza et al., 2018, p. 18). During the debriefing, "team members reflect on a recent experience, discuss what went well, identify opportunities for improvement, and agree on what they will do going forward." Thus, "debriefing promotes learning, enables teams to adjust," (Tannenbaum & Greilich, 2022, p. 1), "allows for team performance examination (Rajwani, 2019, p. 1), "effective, time-

and cost-efficient strategy" (Fatima et al., 2020, p. 415), "facilitates reflective thinking through structured discussion and feedback" (Sahin & Basak, 2021, p. 1), a tool for stress management (Ko & Choi, 2020, p. 11), a method to "determine opportunities for improvement at the individual, team, and system level" (Allam et al., 2020, p. 9), which is "beneficial for quality assurance"

Allen et al. (2018) Succinctly outlined the evidence-based practices for effective debriefing as revealed by Salas et al. (2008). These practices are as follows:

◊ Debriefs must be diagnostic (i.e., identify specific ways to improve work).

◊ Ensure that the organization creates a supportive learning environment for debriefs.

◊ Encourage team leaders/members to be attentive during performance regarding what they may want to discuss later (i.e., work tasks to be debriefed).

◊ Educate team leaders on the science of leading team debriefs (i.e., facilitation processes).

◊ Ensure that team members feel comfortable in debriefs (e.g., psychological safety).

◊ Focus on a few critical performance issues during the debrief (i.e., less is more).

◊ Describe specific teamwork interactions and processes involved in the team performance.

◊ Support feedback with objective data.

◊ Provide outcome feedback later (i.e., not during the debrief) and less frequently than process feedback.

◊ Provide both individual and team-oriented feedback at appropriate times.

◊ Shorten the time delay between task performance and debriefing.

◊ Record conclusions made and goals set during the debriefing and follow-up. (p. 508)

While the debriefing process holds significant value as an intervention tool for team building, the absence of debriefing values can be attributed to "interpersonal factors, such as communication breakdown" (Arriaga et al., 2019, p. 1039). Thus, several critical pitfalls should be diligently avoided during debriefing sessions. Reyes et al. (2018) identified five of these critical pitfalls including the following:

◊ Too much focus on task work; leaders do not naturally bring up teamwork issues, which may be creating some of the task work problems.

◊ Telling, not discussing; leaders do not engage the team in diagnosing team effectiveness and developing action plans.

◊ Improper or inadequate focus; debriefing should have adequate structure, avoid trivial or non-negotiable issues, and help the team move forward by asking a question about a meaningful team issue.

◊ Good look back, but no definitive look forward; there should be a transition from looking backward to looking forward.

◊ Too evaluative or threatening; Research has shown that perceived purpose (judgmental versus developmental) has a significant impact on the accuracy and acceptance of feedback.

The techniques of effective team building we have covered are the following:

◊ Improve workgroup skills.

◊ Develop social relationships.

◊ Solve problems.

◊ Create the necessary conditions for the formation of skills of effective teamwork.

◊ Create an atmosphere of information communications.

◊ Perform team task analysis.

◊ Construct a teamwork design.

◊ Provide team charters.

◊ Offer training interventions.

◊   Provide coaching intervention.

◊   Perform team performance monitoring and assessment.

◊   Perform team debriefing.

Workplace coaching programs aimed at enhancing the work behaviors and well-being of managers provide preliminary evidence in favor of workplace coaching as an effective approach for facilitating work effectiveness. Passmore et al. (2020) found that

> Participation in coaching was associated with significantly enhanced proactivity, core performance, goal-attainment, self-insight, motivation, positive affect, and autonomy. Significant effects on self-reflection, negative affect, and psychological well-being were not found. [Thus] coaching was a more effective and efficient method for learning and facilitating an improved relationship and helped the learner to learn more quickly.

Beyond coaching, effectively managing construction teams

In the practical sense, when employees in a team or a construction project are allowed to advance their knowledge through aspects such as workshops, feedback mechanisms, and the provision of necessary resources, they are more likely to excel. It is essential that the employee's well-being is taken into context as this is the primary enabler of their success. All strategies that encourage the individual and collective development of construction team members are closely related to better productivity, commitment, innovation, collaboration and ultimately the team's success.

Based on these research outcomes, recommendations for fostering effective team development within the construction industry were delineated by the nine experts I interviewed as supported by the survey. These include establishing clear project plans, setting team boundaries, fostering social connections, encouraging positive interactions and proactive problem-solving, arranging technical workshops, leading through collaboration, implementing feedback mechanisms, providing necessary resources, ensuring a safe work environment, conducting regular team meetings, recognizing exceptional performance, establishing protocols, and encouraging unity and empathy within the team. Incorporating these strategies into construction team management can significantly enhance teamwork, productivity, and project outcomes. Customizing these approaches to suit specific team dynamics within the construction industry is crucial for optimal success.

But in this chapter, we are specifically discussing the role of the leader. What are the fundamental leadership characteristics necessary for steering and managing construction teams effectively? My maturity model holds, as I noted above, that team leadership consists of five basic factors: recognition and appreciation, goal setting and excellent communication, continuous improvement and development of the team, encouraging team motivation and employee well-being, and delegation though collaboration and trust-building.

Again, the interviews stressed the role of effective communication, this time specially in leadership. In fact, it quickly became evident that effective communication was the cornerstone of their approach. "We communicate with the team on a regular basis," one member said, setting the tone for the discussion. Others emphasized the importance of speaking openly, using various communication channels effectively, and ensuring that everyone was kept informed and aligned with project objectives. "Make sure everyone has a voice," chimed in another team member. The team members stressed the need to create an environment where every team member felt heard and valued. "Communicating a clear vision and expectations with the team is essential," added another, highlighting the importance of setting a unified direction for the project. The team members spoke passionately about the necessity of active listening, adapting communication styles, and the constant reception and implementation of feedback to fine-tune their communication skills. "Building trust through transparent and open communication is the bedrock of our teamwork," one member shared, underscoring the significance of honesty and transparency.

However, the experts acknowledged that conflicts were inevitable in the high-pressure construction environment. "Accepting conflict among team members is crucial," one participant stated. "But we don't let conflicts fester; we become calming agents," another member explained. They described a process that involved listening actively to all parties involved, analyzing the root causes, and separating the person from the problem. "Encouraging open discussions during disputes helps us recognize differences among members' perspectives," one team member revealed. The interviewers emphasized the importance of finding areas of agreement and disagreement, focusing on behavior and events rather than personalities, and building bridges through brainstorming solutions. Their insights extended beyond conflict resolution. "We promote compromise and cooperation solutions," one team member stated. "Developing action plans is key," another added, highlighting the importance of translating resolutions into tangible actions.

"To ensure the team understands the bigger picture of the project, we discuss obstacles in our weekly meetings," a team member shared, underlining their commitment to maintaining transparency. "Fixing a good working schedule and ensuring clear task understanding is pivotal," another mentioned, highlighting their emphasis on operational excellence. "Establishing a consensus and cooperation environment helps the team function well," one participant explained, elucidating their approach to fostering team cohesion. "We create a supportive context and encourage team collaboration," another added. Getting to know team members and being transparent with them was a recurring theme. "Punctuality and full cooperation with all participants are non-negotiable," one member stated firmly. They stressed the importance of unity, friendliness, and sympathy among team members, listening actively, and creating an environment for sharing ideas and questions. The conversation with one team member turned to the broader team culture. "We foster a positive team culture that encourages collaboration, innovation, and learning," one member shared, emphasizing the need to build trust, responsibility, respect, and mental and physical health.

"To start things off right, we hold an early introduction meeting where we explain the project requirements, and share the clear purpose, goals, objectives, and expectations," another team member explained. They stressed the importance of ensuring that each person fully understood their role and responsibilities. The discussion shifted toward strategies for task management. "We prioritize tasks and provide guidance on task management," one member stated. "Balancing workloads for optimal performance is essential," another added, highlighting the team's focus on efficiency. "Setting SMART goals and objectives, and aligning team efforts toward common goals are part of our daily routine," one participant shared. They spoke of the need to avoid misunderstandings and misinterpretations in task distribution and the importance of defining and documenting team processes and procedures for task execution.

"Clear accountability and providing feedback are vital for performance management," one team member explained. They mentioned the significance of monitoring team development, resolving issues, and continually improving against targets and expectations. The team members offered insights into their strength-based approach. "We align tasks and responsibilities with individual strengths," one participant said, emphasizing the practice of leveraging competency mapping for effective team performance. "Delegating tasks based on strengths is key," added another. The discussion then delved into the importance of using project management software for efficient work organization and ensuring a clear action plan and regular updates. "Keeping the team informed is essential," a team member stated, advocating for transparency. "We promote efficiency and avoid micromanagement," another added, emphasizing their commitment to empowering team members. "Regular tracking and recording of the team's performance are part of our routine," one participant mentioned. They highlighted the need to take action continually to improve against targets and expectations and the importance of conducting annual performance evaluations. Maintaining internal and external auditing was deemed essential for accountability. "Providing feedback and monitoring team development are ongoing processes," a team member explained. The team members reflected on their continuous guidance and direction of team members to achieve their goals. "We identify and understand causes for changes in behaviors to maximize productivity," one member shared. They stressed the importance of discussing key challenges and listening carefully to their team. The interviews revealed a keen focus on team dynamics. "We observe positive or negative outliers in the team," a participant added, explaining their approach to identifying and addressing issues. "We coach our team members to overcome the difficulties they face," another added, emphasizing the importance of supporting each other. "Fostering and encouraging a shared mindset is vital," one team member stated, emphasizing the need for unity and shared vision. They stressed the importance of treating the whole team fairly, without discrimination, and encouraging team members to gain self-confidence.

"Providing adequate training for knowledge and skill development is a continuous effort," one team member explained, highlighting their commitment to growth and improvement. They also mentioned their practice of seeking feedback from teammates and leaders about difficulties they faced. "We trust our team members to commence and continue their duties independently," one member stated, reflecting their belief in the capabilities of their colleagues.

I wanted to explore the strategies and criteria imperative for assembling construction teams in a manner that facilitates effective management and direction. While team composition was an individual aspect of consideration as part of the construction effectiveness model, the research noted that it also came up strongly under the organizational culture as already discussed and that underlines why it was necessary to examine it independently because it straddles across the efforts of a success of an effective construction team. These identified factors collectively formed a framework that serves to enhance the management of construction project teams, ensuring optimal performance and outcomes within the industry.

In general, the interviews painted a vivid picture of a construction team dedicated to effective communication, collaborative conflict resolution, team environment and structure, roles and responsibilities clarity, performance management, and planning and task. Their wisdom offers a roadmap for managing the intricate life cycle of construction teams, ensuring not only the completion of projects but also the fostering of a thriving and harmonious work environment.

The sum of what I took regarding effective management of construction teams involves seven factors, all with multiple techniques. Note that many of these techniques are shared by various factors, but they often differ in purpose depending on the context of those factors, as follows:

◊ Communication

   ◊ Communicate with the team on a regular basis.

   ◊ Speak openly with your team.

   ◊ Use various communication channels effectively.

   ◊ Keep everyone informed and aligned.

   ◊ Make sure everyone has a voice.

   ◊ Communicate clear vision and expectations with the team.

   ◊ Regularly refresh the team's understanding of contract documents and client/developer requirements

   ◊ Actively listen to team members.

   ◊ Adapt communication styles.

   ◊ Receive and implement feedback.

   ◊ Build strategies for improving communication skills.

   ◊ Build trust through transparent and open communication.

   ◊ Share information openly and honestly

◊ Conflict Management and Dispute Resolution

   ◊ Accept conflict among team members.

◊ Be a calming agent.

◊ Listen actively to all parties involved in the conflict.

◊ Analyze and understand the root cause of conflict.

◊ Separate the person from the problem.

◊ Encourage open discussions during disputes.

◊ Recognize differences among members' perspectives.

◊ Find areas of agreement and disagreement.

◊ Focusing on behavior and events rather than personalities.

◊ Building bridges during conflicts.

◊ Brainstorming solutions to resolve the conflicts.

◊ Promoting compromise and cooperation solution.

◊ Developing action plans.

◊ Ensure the team understands the bigger picture of the project.

◊ Discuss in the weekly meetings the obstacles the team faces.

◊ Provide a good working schedule and ensure clear task understanding.

◊ Look beyond the incident of conflict.

◊ Team Environment and Structure

  ◊ Establish a consensus and cooperation environment.

  ◊ Organize the team to start functioning well.

  ◊ Establish clear boundaries and interdependence among members.

  ◊ Establish compelling direction.

  ◊ Enable structure, create a supportive context, and encourage team collaboration.

  ◊ Know the team members, and be transparent with them.

  ◊ Prove punctuality and full cooperation with all participants.

  ◊ Encourage unity, friendliness, and sympathy among team members.

  ◊ Listen to team members

  ◊ Create an environment for sharing ideas and questions.

  ◊ Foster a positive team culture that encourages collaboration, innovation, and learning

  ◊ Build trust, responsibility, respect, and mental and physical well-being.

◊ Goal Setting, Clear Roles, and Expectations

◊   Hold an early introduction meeting and explain the project requirements

◊   Share the clear purpose, goal, objectives, and expectations.

◊   Ensure each person understands their role.

◊   Establish strategies for prioritizing tasks and responsibilities.

◊   Balance workloads for optimal performance.

◊   Set SMART goals and objectives.

◊   Align team efforts toward common goals.

◊   Avoiding misunderstandings and misinterpretations in task distribution.

◊   Arrange weekly meetings with the team at the beginning to make things clear.

◊   Set KPIs to measure progress and establish clear reporting system.

◊   Prioritize tasks and provide guidance on task management.

◊   Emphasizing the importance of focusing on the project schedule.

◊   Define and document team processes and procedures for task execution.

◊ Set clear accountability.

◊ Provide feedback.

◊ Ensure that the team has access to the necessary resources.

◊ Provide awareness of project conditions, government regulations, end users, and external stakeholders.

◊ Align tasks and responsibilities with individual strengths.

◊ Leverage competency mapping for effective team performance.

◊ Planning and Task Distribution

◊ Delegate tasks based on strengths.

◊ Use good project management software to organize the work.

◊ Ensure availability of a clear action plan and its updates.

◊ Clearly define roles and responsibilities for every team member.

◊ Keep the team informed.

◊ Promote efficiency, and avoid micromanagement.

◊ Team Performance Management

◊ Keep regular track of and record the team's performance.

◊ Take action to continually improve against targets and expectations.

◊ Conduct annual performance evaluations for all teams.

◊ Maintain internal and external auditing.

◊ Provide feedback.

◊ Monitor team development and resolve issues.

◊ Continuously guide and direct team members to achieve established goals.

◊ Identify and understand causes for changes in behaviors to maxim productivity.

◊ Discuss key challenges, and listen carefully.

◊ Observe positive and negative outliers in the team.

◊ Coaching and Skills Enhancement

◊ Coach your team members to overcome the difficulties they face.

◊ Foster and encourage the shared mindset.

◊ Treat the whole team fairly and without discrimination.

◊ Encourage the team members to gain self-confidence.

◊ Ensure team members are fully aware of their duties and responsibilities.

◊ Focus on communication, problem-solving, and cooperation.

◊ Provide adequate training for knowledge and skill development.

◊ Seek feedback from teammates and leaders about difficulties they face.

◊ Provide necessary support and resources for success.

◊ Trust the team members to commence and continue their duties independently.

◊ Understand the competency of each team member in solving issues/problems.

◊ Encourage activities that strengthen relationships between members.

This rather unwieldy list was ultimately the material I used in constructing a robust and all-encompassing maturity model specifically tailored for the effective management of construction teams (EMCTM2). I remind the reader of my model's configuration, which is easier to review.

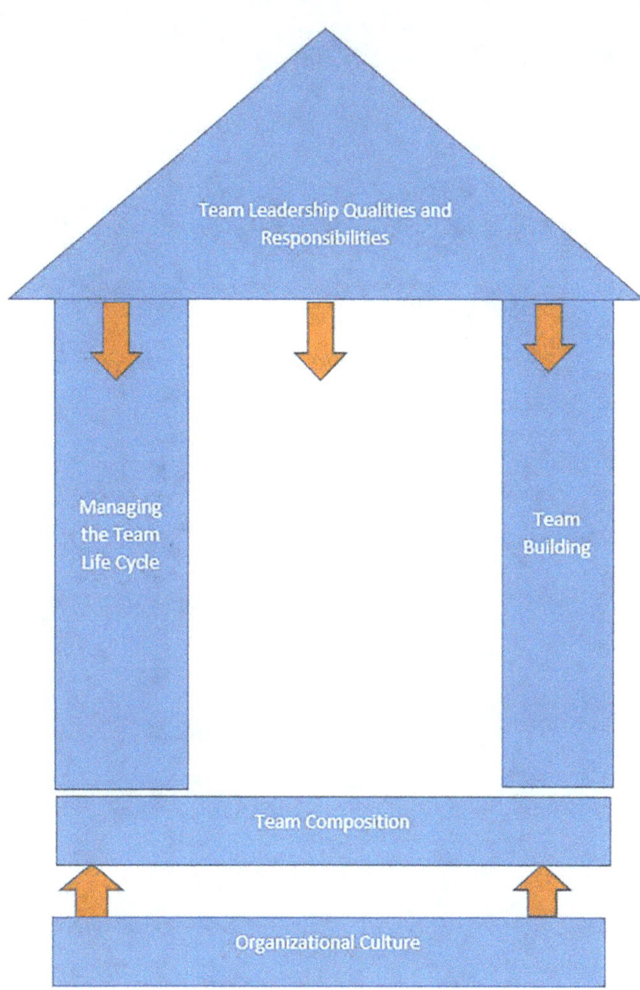

Figure 4

Effective Management of Construction Teams Model

Chapter 10

Conclusion

From a practical perspective, the management of the construction project team is a challenge that every project manager has to navigate. There are a number of strings to join to have the team running effectively, and if these strings are not effectively manipulated, then it is impossible for the team to succeed. Thus, I began work on this book by asking the question, what factors and techniques influence the management of construction teams from the practitioners' perspective?

This question led to the investigation of the issue of management of effective construction teams from a more technical and practical aspect. The overarching question was further subdivided into several sub-questions to comprehensively address all facets relevant to the efficacy of team management within the construction context. These sub-questions encompassed various dimensions identified through the extensive literature review, ensuring a thorough examination of factors crucial for effective construction team management. The aim was to acquire detailed insights and answers pertaining to the optimal strategies and practices necessary for the successful management of construction teams, as informed by the perspectives and experiences of practitioners in the field.

This inquiry took me through well-known territory, such as organizational culture. According to Warrick (2017), "Organizational culture can significantly influence the performance and effectiveness of a company; the morale and productivity of its employees; and its ability to attract, motivate, and retain talented people" (p. 395). The research findings illuminated the pivotal role of organizational culture in shaping the dynamics, functionality, and overall success of construction project teams. These findings underscore the critical significance of organizational culture and specific practices in influencing the effective management of teams within the construction industry.

We then discussed issues related to team formation and composition, highlighting the importance of certain enabling conditions, such as the right mix of individuals, for team success. As Bakker (2017) said, "Scholars in the domain of human resource management have argued that organizations can increase employee work engagement by selecting the candidates who are best suited to the job and fit with the organization's culture" (p. 68). Mingaleva et al. (2022) stressed the importance of team formation and composition strategies, suggesting that organizations identify employees whose personality traits and belief systems align with the desired organizational culture, referring to them as "cultural donors." Pham and Dinh (2020) emphasized the importance of fostering a culture of collaboration, trust, respect, and effective communication, stating that an open culture necessitates creating a friendly environment for sharing, conversation, and collaboration.

The nine experts interviewed addressed these topics but placed them in the real world. In hearing from them, and using quantitative data from the survey to support them, we were able to understand these issues in their practical function, not as philosophies or academic subjects. In this way, this book bridges of the gap in knowledge regarding the management of effective construction teams, as I discussed in the first chapter. In the review of literature on the subject matter, the research found a myriad of impressive research pieces that have been cited herein which dispensed knowledge on construction teams' management and the different factors that lead to the effective management of construction teams. However, all through the different pieces of research, one element was lacking and that is the convergence of these different factors toward the development of an integrated understanding of the process of management of construction teams. The research noted that the different evidences of research investigated some factors while leaving others, while the bulk of them investigated the factors for the purposes of investigating them without a clear goal of how to use the knowledge after investigating and understanding the factors.

By conducting the investigation in the current study with the goal of understanding the tapestry behind the interrelated factors affecting the effective management of construction teams, this work addresses the lack of knowledge surrounding the management of effective construction team's methodologies. I did not want just to identify, yet again, the facto factors and fundamental characteristics impacting the effective management of teams in construction projects but to clearly see how and why the factors should work together.

From a theoretical perspective, the developed maturity model I offer here is not exhaustive or conclusive in nature, rather, it is a starting point for myself and for others to promote a better understanding of the importance of integrating the different factors impacting the effective management of construction teams, moving toward better methodologies of management of construction teams. With a low output in terms of construction team management methodologies as identified in the literature, this book's contributions in the field of construction teams' management methodologies are cutting-edge and open up room for further studies in the field. These studies would then lead to the demystification of the challenges of effective management of construction teams by developing theories and research-based guidelines that can help in promoting the effective management of construction teams.

The book's contributions to practice for the current study are significant. There is no doubt that for the current study, the developed maturity model for implementation in the construction industry toward effective construction team management is critical. However, in presenting the contribution to practice, the research will be more practical and highlight the aspects of the maturity model and how they will directly impact the practical management of construction teams. To start with, the policy level of construction team management will be directly impacted by the maturity model developed. As I found in my research, aspects of policy such as organizational culture, with specific reference to motivation behaviors and ethical work behaviors, as well as rewards and the work environment, need to be optimized to ensure that there is a better culture that can support effective team performance in construction projects. I also found that elements such as team composition and the selection criteria, which overlap through the different fundamental characteristics of effective team management in the construction industry, have to be clearly stipulated in the policy as they stand to significantly affect the effectiveness of construction teams.

Another aspect of practical contribution by this work is its promotion of the role of leadership dynamics and its impact on effective construction teams. The team leader must be able to influence the team members positively, develop a good working environment, and inspire the team members by individually leading from the front in terms of performance and delivery during projects. The maturity model clearly encapsulates this, especially when considered from the lens of the quantitative data wherein the relationship between other fundamental characteristics and team leadership style and quality characteristics are compared. The contribution to practice is that team leaders must adopt a leadership approach that is conducive and that ensures that all members of the team are comfortable and engaged to deliver. I note that this part of the fundamental characteristics and factors affecting the effective management of construction teams can hardly be enshrined in policy, rather, it is a practical quality and the only way to achieve it is through a clear and transparent, objective-oriented team composition and building process. This also helps highlight the importance of the practical contribution of the research touching on team development and building which highlights the importance of aspects of team development and building including coaching and training as far as the effective management of construction teams is concerned.

References

Ali, M., Arbi, A., Raza, B., Malik, Z., & Sheikh, L. (2020). Does managerial coaching effect subordinates' feedback orientation? Mediating role of affective supervisor commitment. *Polish Journal of Management Studies, 22*(2), 9–21. http://doi.org/10.17512/pjms.2020.22.2.01

Allam, T., Ahmed, A., & Ghaly, A. (2020). Effect of debriefing learning strategy on nursing students' knowledge and performance. *IOSR Journal of Nursing and Health Science, 9*(1), 8–18. http://doi.org/10.9790/1959-0901040818

Allen, J., Reiter-Palmon, R., Crowe, J., & Scott, C. (2018). Debriefs: Teams learning from doing in context. *American Psychologist, 73*(4), 504–516. http://doi.org/10.1037/amp0000246

Al-Malki, M., & Juan, W. (2018). Leadership styles and job performance: A literature review. *Journal of International Business Research and Marketing, 3*(3), 40–59. http://doi.org/10.18775/jibrm.1849-8558.2015.33.3004

Azmy, N. (2012). *The role of team effectiveness in construction project teams and project performance.* Graduate These and Dissertation—Iowa State University, 1–136.

Bakker, A. B. (2017). Strategic and proactive approaches to work engagement. *Organizational Dynamics, 46*(2), 67–75. http://doi.org/10.1016/j.orgdyn.2017.04.002

Bell, S., & Brown, S. (2015). Selecting and composing cohesive teams. *Managing Groups and Teams, 17,* 181–209. http://doi.org/10.1108/S1534-085620150000017008

Bell, S., Brown, G., Colaneri, A., & Outland, N. (2018). Team composition and the ABCs of teamwork. *American Psychologist, 73*(4), 349–362. http://doi.org/https://dx.doi.org/10.1037/amp0000305

Bisbey, T., Traylor, A., & Salas, E. (2021). Transforming teams of experts into expert teams: Eight principles of expert team performance. *Journal of Expertise, 4*(2), 190–207. https://www.journalofexpertise.org/articles/volume4_issue2/JoE_4_2_Bisbey_etal.html

Bresman, H., & Zellmer-Bruhn, M. (2012). The structural context of team learning: Effects of organizational and team structure on internal and external learning. *Organization Science, 24*(4), 1120–1139. http://doi.org/10.1287/orsc.1120.0783

Brown, J. D. (2016). Statistics corner: Questions and answers about language testing statistics. CreateSpace Independent Publishing Platform. https://teval.jalt.org/test/PDF/Brown30.pdf

Chidambaram, V., Shanmugam, K., & Sivamani, B. (2021). Effect of project team integration on the performance of Indian construction project: SMART PLS structural equation approach.

International Journal of Construction Supply Chain Management, *11*(1), 1–20. http://doi.org/10.14424/ijcscm110121-1-20

Chin, W. W. (2010). How to write up and report PLS analyses. In *Handbook of partial least squares: Concepts, methods and applications* (pp. 655–690). Springer. http://doi.org/10.1007/978-3-540-32827-8_29

Choi, S., Kim, K., & Kang, S.-W. (2017). Effects of transformational and shared leadership styles on employees' perception of team effectiveness. *Social Behavior and Personality an International Journal, 45*(3), 377–386. http://doi.org/10.2224/sbp.5805

Dansoh, A., Oteng, D., & Frimpong, S. (2017). Challenges associated with project teams in managing variations on construction projects. 6TH International Conference on Infrastructure Development in Africa. 12–14, pp. 121–131. https://d1wqtxts1xzle7.cloudfront.net/54031172/

De Vries, M. (2020). Executive group coaching: Interventions not for the faint of heart. *Advances in Global Leadership, 12,* 153–171. http://doi.org/10.1108/S1535-120320190000012009

Dolgov, M., Dotsenko, E., Nefedov, A., & Zverev, S. (2006). Team management in Russia. Methods of formation and development of team effects in the organization. HR—Portal. https://hr-portal.ru/article/komandnyymenedzhment-v-rossii-metody-formirovaniya-i-razvitiya-komandnyh-effektov-v

Epton, T., Currie, S., & Armitage, C. J. (2017). Unique effects of setting goals on behavior change: Systematic review and meta-analysis. *Journal of Consulting and Clinical Psychology*, *85*(12), 1182–1198. http://doi.org/10.1037/ccp0000260

Faris, H., Gaterell, M., & Hutchinson, D. (2022). Investigating underlying factors of collaboration for construction projects in emerging economies using exploratory factor analysis. *International Journal of Construction Management*, *22*(3), 514–526. http://doi.org/10.1080/15623599.2019.1635758

Fincham, J. (2008). Response rate and responsiveness for surveys, standards, and the journal. *American Journal of Pharmaceutical Education*, *72*(2), 1–3. http://doi.org/10.5688/aj720243

Fitria, H. (2018). The influence of organizational culture and trust through the teacher performance in the private secondary school in Palembang. *International Journal of Scientific & Technology Research* , 7(7), 82–86. https://www.researchgate.net/publication/329884811_The_Influence_Of_Organizational _ Culture_And_Trust_Through_The_Teacher_Performance_In_ The_Private_Secondary_ School_In_Palembang

Fokina, O., & Krupnov, Y. (2020). Evaluation of the company's organizational culture. *Innovative Technologies in Science and Education*, *210*, pp. 1–6. http://doi.org/10.1051/e3sconf/202021010007

Fong, P., & Kwok, C. (2009). Organizational culture and knowledge management success at project and organizational levels in contracting firms. *Journal of Construction Engineering and Management*, *12*(135), 1348–1350.

Fontannaz, S., & Cox, E. (2020). Team leadership development through coaching: A case study of skippers in a round-the-world sailing race. *International Journal of Evidence Based Coaching and Mentoring, 18*(2), 19–34. http://doi.org/10.24384/14k4-pa14

Gamil, Y., & Abdul Rahman, I. (2017). Identification of causes and effects of poor communication in construction industry: A theoretical review. *Emerging Science Journal, 1*(4), 239–247. http://doi.org/10.28991/ijse-01121

Gilal, A., Jaafar, J., Capretz, L., Omar, M, Basri, S., & Aziz, I. (2017). Finding an effective classification technique to develop a software team composition model. *Journal of Software: Evolution and Process, 10*(29), 1–22. http://doi.org/10.1002/smr.1920

Ginting, H., Mahiranissa, A., Bekti, R., & Febriansyah, H. (2020). The effect of outing team building training on soft skills among MBA students. *The International Journal of Management Education, 18*, 1–12. http://doi.org/10.1016/j.ijme.2020.100423

Gren, L., Torkar, R., & Feldt, R. (2016). Group development and group maturity when building agile teams: A qualitative and quantitative investigation at eight large companies. Journal of Systems and Software, 124, 1–32. http://doi.org/10.1016/j.jss.2016.11.024

Grynchenko, M., Ponomayov, O., & Lobach, O. (2018). Leadership as factor for building a project team. *Innovative Technologies and Scientific Solutions for Industries, 1*(3), 13–21. http://doi.org/10.30837/2522-9818.2018.3.013

Guilherme, M. M., da Cruz, A. P., & Barbosa, M. A. (2019). The role of human performance assessment in a public educational institute. *ConTexto—Contabilidade em Texto, 19*(41), 29–42. https://seer.ufrgs.br/index.php/ConTexto/article/view/88702

Hair, J. F., Howard, C. M., & Nitzl, C. (2020). Assessing measurement model quality in PLS SEM using confirmatory composite analysis. Journal of Business Research, 109, 101110. http://doi.org/10.1016/j.jbusres.2019.11.069

Islami, X., Mulolli, E., & Mustafa, N. (2018). Using management by objectives as a performance appraisal tool for employee satisfaction. *Future Business Journal*, *4*, 94–108. http://doi.org/10.1016/j.fbj.2018.01.001

Ivanov, S., McFadden, M., & Anyu, J. (2021). Examining and comparing good and bad leaders based on key leadership characteristics: A leadership case study. *International Journal of Organizational Innovation*, *13*(3). https://www.researchgate.net/publication/351103131_examining_and_comparing_good_and_bad_leaders_based_on_key_leadership_character istics_a_leadership_case_study

Jena, L. K., Pradhan, S., & Panigrahy, N. P. (2018). Pursuit of organisational trust: Role of employee engagement, psychological well-being and transformational leadership. *Asia Pacific Management Review*, *23*, 227–234. http://doi.org/10.1016/j.apmrv.2017.11.001

Johnson, R., & Babu, R. (2018). Time and cost overruns in the UAE construction industry: A critical analysis. *International Journal of Construction Management, 20*(5), 402–411. http://doi.org/10.1080/15623599.2018.1484864

Jones, A. (2019). The Tuvkman's Model Implementation, Effect, and, Analysis & The New Development of Jones Lis Model on

References

A Small Group. *Journal of Management, 6*(4), 2338. http://doi.org/10.34218/JOM.6.4.2019.005

Jones, R., Napiersky, U., & Lyubovnikova, J. (2019). Conceptualizing The distinctiveness of team coaching. *Journal of Managerial Psychology, 34*(2), 62-78. http://doi.org/10.1108/JMP-072018-0326

Katane, K., & Dube, S. (2017). The influence of organizational culture and project management maturity in virtual project teams. 2nd International Conference on the Internet, Cybersecurity and Information Systems (ICICIS 2017). 1, pp. 1–9. Johannesburg. https://www.researchgate.net/publication/319528141

Keck, S. L., & Tushman. (2017). Environmental and organizational context and executive team structure. *Academy of Management Journal, 36*(6). https://journals.aom.org/doi/epub/10.5465/256813

Ku, O., Liang, J., Change, S., & WU, M. (2019). A preliminary case study of promoting teacher professional development via post-lesson debriefing with the support of Sokrates cloud services. Proceedings of the 27th International Conference on Computers in Education (pp. 1-7). Taiwan: Asia-Pacific Society for Computers in Education. https://apsce.net/icce/icce2019/proceedings/paper_168.pdf

Kuzior, A., & Balahurovska, I. (2023). Interaction between the manager and the hidden informal leadership. *Organization and Management Series,* 173-181. http://doi.org/10.29119/1641-3466.2022.165.12

Lacerenza, C., Marlow, S., Tannenbaum, S., & Salas, E. (2018). Team development interventions: Evidence-based approaches for improving teamwork. *American Psychologist, 73*(4), 517–531. http://doi.org/10.1037/amp0000295

Latif, F., & Williams, N. (2017). Team effectiveness in non-governmental organizations (NGOs) projects. *Evaluation and Program Planning, 64,* 20–32. http://doi.org/10.1016/j.evalprogplan.2017.05.004

Lee, D. (2020). Impact of organizational culture and capabilities on employee commitment to ethical behavior in the healthcare sector. *Service Business, 14,* 47–72. http://doi.org/https://doi.org/10.1007/s11628-019-00410-8

Liu, D., Wang, H. -w., Li, H., Wang, J., & Khallaf, M. (2019). Hierarchical task network approach for time and budget constrained construction project planning. *Technological and Economic Development of Economy, 25*(3), 472-495. http://doi.org/10.3846/tede.2019.9384

Loberg, L., Nüesch, S., & Foege, J. N. (2018). Forced distribution rating systems and team performance: The moderating role of task interdependence. Academy of Management, 2018(1). http://doi.org/10.5465/AMBPP.2018.12355abstract

Luo, L., Yang, Y., Wu, G., Zheng, J., & Liu, D. (2023). Effects of organizational leadership on project citizenship behavior and management performance in complex construction projects. *Buildings, 13*(1), 1–20. http://doi.org/10.3390/buildings13010259

Lyons, R., Lazzara, E., Benishek, L., & Zajac, S. (2015). Enhancing the effectiveness of team debriefings in medical simulation: More best practices. *The Joint Commission Journal on Quality and Patient Safet, 41*(3). http://doi.org/10.1016/S1553-7250(15)41016-5

Marks, M., Matheu, J., & Zaccaro, S. (2001). A temporally based framework and taxonomy of team processes. *Academy of Management Review, 26*(3), 356–376. https://www.jstor.org/stable/259182

Mathieu, J., Hollenbeck, J., Knippenberg, D., & Ilgen, D. (2017). A century of work teams in the Journal of Applied Psychology. *Journal of Applied Psychology, 102*(3), 452–467. http://doi.org/https://doi.org/10.1037/apl0000128

McEwan, D., Ruissen, G., Eys, M., Zumbo, B., & Beaucham, R. (2017). The effectiveness of teamwork training on teamwork behaviors and team performance: A systematic review and meta-analysis of controlled interventions. PLos One, 12(1), 1–23. http://doi.org/10.1371/journal.pone.0169604

McLachlan, G.J. Mahalanobis distance. *Reson* **4**, 20—26 (1999). https://doi.org/10.1007/BF02834632

Merriam, S. B. (1998). Qualitative research and case study applications in education. In S. B. Merriam, *Case study research in education* (pp. 27–43). Jossey-Boss. http://eds-courses.ucsd.edu/tep288a/shortbook.pdf

Mingaleva, Z., Shironina, E., Lobova, E., Olenev, V., Plyusnina, L., & Oborina, A. (2022). Organizational culture management as an element of innovative and sustainable development of enterprises. *Sustainability, 14*(6289), 1–28. http://doi.org/https:// doi.org/10.3390/su14106289

Mohsen, A., Neyazi, N., & Ebtekar, S. (2020). The impact of organizational culture on employees' performance: An overview. *International Journal of Management, 11*(8), 879–888. http://doi. org/10.34218/IJM.11.8.2020.079

Moura, I., Dominguez, C., & Varajão, J. (2021). Information systems project team members: Factors for high performance. *The TQM Journal, 33*(6), 1426–1446. http://doi.org/10.1108/ TQM07-2020-0170

Natvig, D., & Stark, N. (2016). A project team analysis using Tuckman's model of small group development. *Journal of Nursing Education, 55*(12), 675–681. http://doi.org/10.3928/01484834- 20161114-03

Neto, A. L., Jr., O. S., & Da Silva, L. F. (2023). The importance of soft skills in the selection of project teams in a company in the agile transformation stage. *Journal on Innovation and Sustainability, 14*(1), 4-18. http://doi.org/10.23925/2179- 3565.2023v14i1p4-18

Ozili, P. K. (2023). The acceptable R-Square in empirical modelling for social research. *MPRA Munich Personal RePEc Archive*, 1–9. http://doi.org/10.4018/978-1-6684-6859-3.ch009

Paolucci, N., Dimas, I., Zappalà, S., Lourenço, P., & Rebelo, T. (2018). Transformational leadership and team effectiveness: The mediating role of affective team commitment. *Journal of*

Work and Organizational Psychology, 34(3), 135–144. http://doi.org/10.5093/jwop2018a16

Passmore, J., & Gordon, S. (2018). International coaching psychology review. *The Australian Psychological Society, 13*(1), 1–112. https://organisationalpsychology.nz/wp-content/uploads/2019/07/Volume_13_No_2_Autumn_2018.pdf

Passmore, J., & Rehman, H. (2020). Coaching as a learning methodology. In D. Tee, & J. Passmore, Coaching researched: A coaching psychology reader. http://doi.org/10.1002/9781119656913.ch18

Passmore, J., van Nieuwerburgh, C., & Barr, M. (2019). *Workplace coaching*. Oxford University. http://centaur.reading.ac.uk/83129/

Pathiranage, Y., Jayatilake, L., & Abeysekera, R. (2020). A literature review on organizational culture towards corporate performance. *International Journal of Management, Accounting and Economics, 7*(9), 522-544. https://www.researchgate.net/publication/344667769_A_Literature_Review_on_Organizational_Culture_towards_Corporate_Performance

Payne, A., & Harper, C. (2020). Studying the impact of personality traits on team performance for construction and engineering. *Korea Institute of Construction Engineering and Management., 10*(3-4), 1–24. http://doi.org/https://doi.org/10.6106/JCEPM.2020.10.3-4.001

Peters, J. and Carr, C. (2013a) 'Team effectiveness and team coaching literature review', Coaching: An International Journal of Theory, Research and Practice, 6(2), pp.116-136. DOI: 10.1080/17521882.2013.798669.

Pham, Q. T., & Dinh, X. K. (2020). The impact of organizational culture on knowledge transferring in FDI enterprises in VIETNAM. *International Journal of Innovation*, *8*(3), 392–411. http://doi.org/10.5585/iji.v8i3.16059

Piwowar-Sulej, K. (2021). Organizational culture and project management methodology: research in the financial industry. *International Journal of Managing*, *4*(6), 1270–1289. http://doi. org/10.1108/IJMPB-08-2020-0252

Putra, A., Hariani, M., Nurmalasari, D., Irfan, M., & Al Hakim, Y. (2020). Role of work environment and organizational culture to job performance. *Journal of Islamic Economic Perspective*, *1*(2). https://jurnalfebi.iainjember.ac.id/index.php/JIEP/article/ view/20

Rahman, R., Radzi, A., Saad, M., & Doh, S. (2020). Factors affecting the success of highway construction projects: the case of Malaysia. *Materials Science and Engineering*, *7*(12), 1–8. http:// doi.org/10.1088/1757-899X/712/1/012030

Rahmani, F., Scott-Young, C., Tadayon, A., & Walt, J. (2021). Team composition in relational contracting (rc) in large infrastructure projects: A Belbin's team roles model approach. *Engineering Construction & Architectural Management*, 1–20. http://doi.org10.1108/ECAM-11-2020-0941

Rajwani, K. (2019). Debriefing in critical care. *Qatar Medical Journal*, *2019*(2), 1–2. http://doi.org/10.5339/qmj.2019.qccc.43

Raza, B, Ali, M, Ahmed, S, & Ahmad, J. (2018). Impact of managerial coaching on organizational citizenship behavior:

The mediation and moderation model. *International Journal of Organizational Leadership*, 27–46.

Reyes, D., Tannenbaum, S., & Salas, E. (2018). Team development: The power of debriefing. *People & Strategy, 14*(2), 46–52. https://scholar.google.ae/scholar?hl=en&as_sdt=0%2C5&as_vis=1&q=Team+Developm ent%3A+The+Power+of+Debriefing&btnG=

Romão, S., Ribeiro, N., Gomes, D., & Singh, S. (2022). The impact of leaders' coaching skills on employees' happiness and turnover intention. *Administrative Sciences, 12*(84), 1–15. http://doi.org/10.3390/admsci12030084

Ryu, E. (2011). Effects of skewness and kurtosis on normal-theory based maximum likelihood test statistic in multilevel structural equation modeling. *Behavior Research Methods, 43*(4), 1066–1074. http://doi.org/10.3758/s13428-011-0115-7

Sahin, G., & Basak, T. (2021). Debriefing methods in simulation-based education. *Journal of Education in Nursing, 18*(3), 1–6. http://doi.org/10.5152/jern.2021.57431

Sakalyuk, O. (2021). Models of team building in the personnel management system. *Open Journal Systems, 2*(83), 136–140. http://doi.org/10.35432/1993-8330appa2832021237292

Salas, E., DiazGranados, D., Klein, C., Burke, S., & StagI, K. (2008). Does team training improve team performance? A meta-analysis. *Human Factors and Ergonomics Society, 50*(6), 903-933. http://doi.org/10.1518/001872008X375009

Salas, E., Reyes, D. L., & McDaniel, S. H. (2018). The science of teamwork: Progress, reflections, and the road ahead. *American Psychologist, 73*(4), 593–600. http://doi.org/10.1037/amp0000334

Salihovic, K. (2021). Team coaching in the workplace. A literature review on team coaching and solving performance deficiency in the workplace. ResearchGate, 1-22. http://doi.org/10.13140/RG.2.2.23925.91366

Santos, P. M., & Cirillo, M. A. (2021). Construction of the average variance extracted index for construct validation in structural equation models with adaptive regressions. *Communication in Statistics—Simulation and Computation, 52*(12), 1–13. http://doi.org/10.1080/03610918.2021.1888122

Sarstedt, M., Ringle, C., & Hair, J. F. (2021). *Partial least squares structural equation modeling*. Springer.

Sharma, P. (2017). How coaching adds value in organisations—The role of individual level outcomes. *International Journal of Evidence Based Coaching and Mentoring, 11*, 1–17. https://radar.brookes.ac.uk/radar/file/9c1a02de-7e38-4379-b8c1-4441a056f176/1/special11-paper-01.pdf

Shuffler, M., Diazgranados, D., Maynard, M., & Salas, E. (2018). Developing, sustaining, and maximizing team effectiveness: An integrative, dynamic perspective of team development interventions. *Academy of Management Annals, 12*(2), 1–37. http://doi.org/https://doi.org/10.5465/annals.2016.0045

Silva, L. H., Ghedine, T., Pereira, C. M., Lino, S. R., & Tutida, A. Y. (2021). A management instrument for team performance evaluation. *Revista Produção E Desenvolvimento, 7*, 1–16. http://doi.org/10.32358/rpd.2021.v7.511

Smith, T., Fowler-Davis, S., Nancarrow, S., Ariss, S., & Enderby, P. (2020). Towards a theoretical framework for integrated team leadership! *Journal of Interprofessional Care, 34*(6), 726–736. http://doi.org/10.1080/13561820.2019.1676209

Smolska, M. (2021). A team development process based on the high performance team coaching model: a case study of team maturity management. *Scientific Journals of the Maritime University of Szczecin, 67*(139), 1–8. http://doi.org/10.17402/479

Smolska, M. (2022). The art to select the appropriate development methods for a project team: A case study. *European Research Studies Journal, 25*(1), 527–537.

Ssenyange, K., & Kudakwashe, C. (2023). Impact of leadership styles on government construction project success: A structural equation modelling approach. *Journal of Construction Business and Management, 6*(2), 11–26. http://doi.org/10.15641/jcbm.6.2.1348

Strode, D., Dingsøyr, T., & Lindsjorn, Y. (2022). A teamwork effectiveness model for agile software development. *Empirical Software Engineering, 27*(2), 1–50. http://doi.org/10.1007/s10664-021-10115-0

Tahmasebinia, F., & Song, V. (2022). Significant factors causing delay in the Cambodian construction industry. *Sustainability, 14*(6), 1–24. http://doi.org/10.3390/su14063521

Tannenbaum, S., & Greilich, P. (2022). The debrief imperative: Building teaming competencies and team effectiveness. *BMJ Quality & Safety*, 1–4. http://doi.org/10.1136/bmjqs2022-015259

Turner, J., Baker, R., Ali, Z., & Thurlow, N. (2020). A new multiteam system (MTS) effectiveness model. *Systems, 8*(12), 1–19. http://doi.org/10.3390/systems8020012

Umar, A., Tamsah, H., Mattalatta, M., Baharuddin, B., & Andul Latief, R. (2020). Training effectiveness and team-performance in public organizations. *Journal of Asian Finance, Economics and Business, 7*(11), 1021–1031. http://doi.org/10.13106/jafeb.2020.vol7

Vaida, S., & Serban, D. (2021). Group development stages. A brief comparative analysis of various models. *Studia Universitatis Babeș-Bolyai Psychologia-Paedagogia, 66*(1). http://doi.org/10.24193/subbpsyped.2021.1.05

Warrick, D. D. (2017). What leaders need to know about organizational culture. *Business Horizons, 60*(3), 395–404. http://doi.org/10.1016/j.bushor.2017.01.011

Widdowson, L., Rochester, L., Barbour, P., & Hullinger, A. (2020). Bridging the team coaching competency gap: A review of the literature. *International Journal of Evidence Based Coaching and Mentoring, 18*(2), 35–50. http://doi.org/10.24384/z9zb-hj74

Woo, H. (2017). Exploratory study examining the joint impacts of mentoring and managerial coaching on organizational commitment. *Sustainability, 9*(2), 1–15. http://doi.org/10.3390/su9020181

Yu, N., Collins, C. G., Cavanagh, M., White, K., & Fairbrother, G. (2020). Positive Coaching

Zainal, D., Razali, R., & Mansor, Z. (2020). Team formation for agile software development: A review. *International Journal on Advanced Science Engineering Information Technology, 10*(2), 555–561. https://core.ac.uk/download/pdf/296919946.pdf

Zubanov, V., Katic, I., Grubic-Nesi, L., & Berber, N. (2017). The role of management teams in business success: Evidence from Serbia. *Engineering Economics, 28*(1), 68–78. http://doi.org/10.5755/j01.ee.28.1.15132